PRAISE FOR
Achieving Fluency: Special Educa...

What educators have not said to themselves, "I know that all means all, but how can I do that?" What mathematics teacher has not struggled to find just that right strategy to help a student with special needs? What administrators have not struggled to help their teachers meet every student's need? This book offers educators thought-provoking considerations for how to make "all means all" a reality. When we meet Aaron in chapter 2, we recognize that student who has slipped through the cracks as teachers struggle to maximize his mathematical proficiency. Later chapters walk us through the intricacies of finding the right key to unlock students' learning potential. From using assessments effectively to organizing curriculum to more efficiently develop computational fluency to the conceptual development with all students of algebraic, geometric, measurement, and data analysis and probability topics, this book will become a handbook for all teachers (general, inclusion, or special educators) and administrators (school based or central office) interested in helping all students learn mathematics.

Patricia Baltzley
Director, Mathematics Pre-K–12
Baltimore County Public Schools
Baltimore, Maryland

My teachers often seek my advice as a math resource teacher and coach regarding students who struggle to learn mathematics. These teachers want information, and they don't want to look at twenty different sources to get it. This book is an "all in one," giving both general and special educators a condensed, concise best-practices manual for mathematics instruction. The book addresses content and pedagogical knowledge, as well as an overall understanding of the teaching and learning process for mathematics instruction, including planning, teaching, assessing, and responding to students' needs. Furthermore, the book offers recommendations, guidelines, instructional strategies, and approaches for students needing assistance as well as those designated as special education students. The book also speaks to the idea of, and the components of, mathematical proficiency and then supplies a roadmap to help achieve this goal for all students in the classroom. Using this book will give you the tools to develop a purposeful and thoughtful plan of math instruction while considering the specific needs of your students. You will reach for this reference book throughout the year as you plan your instruction.

Heather C. Dyer
Math Support Teacher
Running Brook Elementary School
Columbia, Maryland

The future of RTI (response to intervention) depends on serious substantive dialogue between mathematics education and special education—a dialogue that has never taken place. This volume offers essential information for this dialogue to begin. Rich with examples, it is an excellent venue for introducing concepts from mathematics education to special educators, and it goes all the way up through algebra and geometry and probability and statistics—areas that the special education area has long neglected.

Russell Gersten
Director, Instructional Research Group
Professor Emeritus, University of Oregon College of Education
Los Alamitos, California

Among students with mathematical difficulties, learning challenges may result from deficient cognitive skills, poor learning opportunities, or both. In each case—as this book demonstrates—high-quality teaching is essential for getting and keeping these students on the path toward mathematics success. Aligned with mathematics standards and principles, this integrated synthesis of research, practice, policy, and standards promotes quality instruction by arming teachers with the knowledge, skills, and tools they need to foster mathematics proficiency. It is an authoritative resource on how to develop and implement effective informal assessments to guide instructional plans and goals, how to determine reasonable learning expectations for children who struggle with math, and how to establish curriculum and content priorities to promote mathematics proficiency among all students—including children with mathematical difficulties. This book is a valuable resource for pre-K–8 educators, most of whom encounter students who struggle with mathematics. As a text, this book is will be extremely useful for courses on special education, mathematics education, and school psychology.

Michèle M. M. Mazzocco
Professor, Psychiatry and Behavioral Sciences, Johns Hopkins University School of Medicine
Principal Investigator, Math Skills Development Project, Kennedy Krieger Institute
Baltimore, Maryland

Achieving Fluency:
Special Education and Mathematics

Edited by
Francis (Skip) Fennell
McDaniel College
Westminster, Maryland

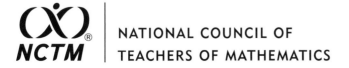

NCTM

NATIONAL COUNCIL OF
TEACHERS OF MATHEMATICS

Copyright © 2011 by
The National Council of Teachers of Mathematics, Inc.
1906 Association Drive, Reston, VA 20191-1502
(703) 620-9840; (800) 235-7566; www.nctm.org
All rights reserved
Second printing 2012

Library of Congress Cataloging-in-Publication Data

Achieving fluency : special education and mathematics / edited by
Francis (Skip) Fennell.
    p. cm.
  Includes index.
  ISBN 978-0-87353-654-7
  1. Mathematics—Study and teaching—United States. 2. Mathematical
ability—Testing. 3. Children with disabilities—Education. 4. Special
education—Curricula—United States. I. Fennell, Francis M., 1944– II.
National Council of Teachers of Mathematics.
  QA13.A338 2011
  510.71—dc22
                     2010054498

The National Council of Teachers of Mathematics is a public voice of mathematics
education, supporting teachers to ensure equitable mathematics learning of the
highest quality for all students through vision, leadership, professional development,
and research.

Printed in the United States of America

Contents

Foreword

"Is it a learning disability or a teaching disability?" This crucial question faces educators when they confront concerns about student performance in mathematics. This issue is a significant one that teachers must examine as they consider how K–12 students with disabilities are developing mathematical knowledge.

The changes in how students with disabilities are included in regular classrooms for mathematics instruction, and in such prevention-based models as response to intervention (RTI), have been dramatic. RTI is based on the assumption that all students receive high-quality mathematics instruction and a high level of assistance when they do not do well. On the basis of a progression from a high-quality curriculum followed by targeted interventions, teachers are expected to move students in focused ways to greater success with mathematical ideas. However, pinpointing these interventions and the related diagnostic assessments is a daunting challenge. Unlike the area of reading, where a stream of research supports the use of particular approaches, mathematics interventions that successfully promote mathematics understanding are neither well researched nor easily found as professional development options.

Achieving Fluency: Special Education and Mathematics presents the understandings that all teachers of mathematics need to play a role in the education of students who struggle: those with disabilities and others who simply received weak instruction in the past or—for whatever reasons—lack essential foundational mathematics knowledge to succeed. This book offers a way for us to delve into what has previously been a silent topic—one that educators and researchers in mathematics education do not often explore. As you read the chapters, you will note the explicit and systematic mathematics instruction that will capitalize on students' strengths and offer the flexibility of thinking to develop mathematically literate citizens. This resource, rather than just a collection of activities, should be by the side of all teachers of mathematics as a foundation for thinking deeply about teaching and learning for students who often do not receive the needed level of support for their success. (See the discussion questions in chapters 5–9.)

Achieving Fluency: Special Education and Mathematics serves teachers and supervisors by sharing increasingly intensive instructional interventions for students who struggle with mathematics on the essential topics aligned with the National Council of Teachers of Mathematics Curriculum Focal Points, the new *Common Core State Standards*, and the mathematical practices and processes that overlap the content. These approaches are particularly useful when initial interventions prove ineffective. For a preventive approach, teachers can align these strategic instructional tools with individual student needs by using a variety of formative assessments.

Too often, educators connect the approaches illustrated in this text only with students with disabilities. But these tools can also be particularly powerful when strategically applied to all students in the ongoing process of developing mathematics knowledge.

Effective teachers work daily to help students develop mathematical understandings. *Achieving Fluency: Special Education and Mathematics* can help with the often-challenging work of increasing essential content knowledge by furnishing structure and guidance for implementing appropriate, effective interventions for all learners.

Karen Karp
University of Louisville, Louisville, Kentucky

CHAPTER 1

All Means All

Francis (Skip) Fennell

History of IDEA

1975: The Education for All Handicapped Children Act (EAHCA) became law. In 1990, Congress amended EAHCA and renamed it the Individuals with Disabilities Education Act (IDEA).

1990: IDEA first came into being on October 30, when Congress amended and renamed EAHCA.

1997: IDEA received several significant amendments. The revision expanded the definition of "disabled children" to include developmentally delayed children aged between three and nine years. The law also required parents to attempt to resolve disputes with schools and local educational agencies through mediation and created a process for doing so. The amendments authorized additional grants for technology, disabled infants and toddlers, parent training, and professional development.

2004: On December 3, Congress amended IDEA, creating the Individuals with Disabilities Education Improvement Act of 2004 (IDEIA 2004). Several provisions aligned IDEA with the No Child Left Behind Act of 2001. The law revised the requirements for evaluating children with learning disabilities. The revision included more concrete provisions relating to discipline of special education students.

2009: President Obama signed the American Recovery and Reinvestment Act (ARRA) on February 17, which included $12.2 billion in additional funds for Services to Children and Youths with Disabilities.

"All means all." This concise phrase has become somewhat of a rallying cry for curricular and instructional opportunities for learners with diverse backgrounds. Such backgrounds most typically imply issues of gender, race, and ethnicity. But perhaps more than at any other time in the history of the United States, and the world, diversity and equity must include opportunities for all children, including those identified as special education students, regardless of their specific learning needs.

In the educational careers of many readers of this resource book, recollections of misguided attempts to provide special education opportunities will be deep-seated memories and, thankfully, no longer commonplace. But these memories will probably include some of the following stereotypical realities related to special education programs, facilities, and teaching:

◆ Classes held in the school basement

◆ Outdated curricular materials

◆ Limited support for special education teachers

◆ Curriculum opportunities and expectations that were, for the most part, skill based only

◆ Significant teacher turnover and a shortage of qualified special education teachers

◆ Limited background related to special education of teachers at the pre-K–8 levels

The Individuals with Disabilities Education Act (IDEA) governs how U.S. states offer special education services to children with disabilities. It addresses the educational needs of children with disabilities from birth to age 21 and involves more than a dozen specific categories of disability. Congress has reauthorized and amended IDEA several times, most recently in December 2004. This book will not explore particular teaching and learning strategies for the various categories of disability, nor will it examine all the intricate challenges related to special education and mathematics. It will, however, address generically the challenges of special education and how teaching—good teaching of important mathematics—is, at minimum, a key to mathematical proficiency.

Although historically, students with disabilities have not had the same access to the general education curriculum as their peers, IDEA has changed the access and accountability requirements for special education students immeasurably. That said, the challenges for meeting the needs of students with disabilities and ensuring their mathematical proficiency confront teachers of mathematics every day.

IDEIA (2004) includes the following instruction-related provisions:

◆ Individualized education program (IEP)—IDEA requires that public schools create and provide an IEP for all eligible students. The IEP is the planning guide for a student's educational program. It specifies the services to be offered and how often, describes the student's present levels of achievement and how the student's disabilities affect academic performance, and specifies accommodations and modifications that the student should receive. The IEP must meet the unique educational needs of each child in the child's least-restrictive environment. The U.S. Department of Education 2005a regulations implementing IDEA state that "to the maximum extent appropriate, children with disabilities, including children in public or private institutions or care facilities, are educated with children who are nondisabled." Basically, the least-restrictive environment is the environment most like that of typical children in which the child with a disability can succeed academically (as assessed by the goals in the student's IEP).

◆ IDEA and No Child Left Behind/Elementary and Secondary Education Act (NCLB/ESEA)—The 2004 reauthorization of IDEA revised the law to align with the requirements of NCLB. NCLB/ESEA allows financial incentives to states that improve their special education services and services for all students. States that do not improve must refund these incentives to the federal government, allow parents choice of schools for their children, and abide by other provisions. Some states are still reluctant to educate special education students and are seeking remedies through the courts. However, IDEA and NCLB/ESEA continue to be the laws of the land. (At this writing, Congress is reauthorizing NCLB. The reauthorization will include restoring the law to its original title as the Elementary and Secondary Education Act.)

The annual adequate yearly progress (AYP) goal of NCLB/ESEA is for increasingly higher percentages of students, all students, to score at the proficient or advanced level on state assessments. However, for special education students to meet such standards, including AYP expectations related to the forthcoming assessments based on the *Common Core State Standards* (Council of Chief State School Officers [CCSSO] 2010), their instruction must incorporate planning, instructional supports, and accommodations, which will include the following:

◆ Special education supports and related services designed to meet the unique needs of special education students and enable access to the general education curriculum (IDEIA 34 CFR §300.34, 2004).

◆ An IEP that includes annual goals aligned with and chosen to facilitate student attainment of grade-level academic standards (state curricular expectations aligned with the *Common Core State Standards).*

◆ Special education teachers and specialized instructional support personnel who are prepared and qualified to deliver high-quality, evidence-based, individualized instruction and support services.

◆ Instructional accommodations that include change in instructional materials or procedures, thus allowing students access to the expectations of state or Common Core state mathematics standards.

◆ Assistive technology and related technological services to ensure student access to skills and concepts appropriate to their needs.

◆ Recognition that for students with significant cognitive disabilities to access certain curricular expectations, those expectations (such as this example from the third-grade *Common Core State Standards* (CCSSO 2010): "Recognize and generate simple equivalent fractions, e.g., $\frac{1}{2} = \frac{2}{4}$, $\frac{4}{6} = \frac{2}{3}$. Explain why the fractions are equivalent, e.g., by using a visual fraction model") may need to be extended or adjusted. Such accommodations should be considered and implemented only after students receive multiple opportunities for learning and demonstrating their knowledge; the accommodations must align with the expectations of the original state or Common Core standards.

What now? Well, for one thing IDEA paired with the accountability expectations of NCLB/ESEA presents true instructional challenges for many teachers. Given these challenges, one could argue that this book is either way overdue or perhaps right on time. But here is what this book is *not*. This professional resource is not a "quick fix, try this today" kind of resource. It is not a collection of neat ideas and activities with nowhere to go. It is not a host of worksheets to plug into IEPs for tomorrow or next week. Rather, this book is the result of years of analyzing research and best practice to give educators a professional resource. This tool will make them think carefully about the challenges they face daily as they strive for an appropriately challenging mathematics experience for their special education students and for those students needing assistance that may range from serious to occasional intervention on particular mathematics topics. The book is also about important mathematics and how to teach it—with specific chapters related to number, algebra, geometry, measurement, and data analysis and probability. This book is your professional resource for thinking about and then conceptualizing instruction for those students needing assistance as well as those designated as special education students.

This chapter forges the policy and school-based framework for *Achieving Fluency: Special Education and Mathematics*. Issues related to IDEA and the daily challenges related to NCLB/ESEA have been established. Let's consider some others.

Response to Intervention

Intervention and the response to intervention (RTI) have become interrelated topics of importance for teachers of mathematics, particularly at the pre-K–8 levels. Where did this concept come from? NCLB/ESEA and IDEA both require school districts to offer support for students who may be having difficulty keeping up with day-to-day expectations, thus establishing the need for intervention. Think of RTI as an early-detection, prevention, and ongoing support system that identifies students and gives them the support they need *before* they fall behind and before they are formally identified and designated for special education services. IDEIA (2004) encouraged states and school districts to use RTI to identify students with learning disabilities and encouraged them to facilitate additional supports for students with academic needs regardless of disability classification. Although many states have begun to implement RTI programs for reading, such initiatives for mathematics are far less common (Gersten, Baker, and Chard 2006).

In essence, intervention is about giving students the opportunity to learn—*all* students—and is a plan for furnishing instructional materials and activities to support student learning during class time, in programs before or after school, and for use by providers of supplemental services.

Intervention programs typically engage a teaching–learning cycle that includes a diagnostic assessment to screen for and identify specific learner needs, instructional actions that are linked to the identified needs, and follow-up assessments to monitor progress. With the impact of NCLB/ESEA and IDEA, it is encouraging that although established RTI programs for mathematics are not as commonplace as they should be, such programs seem to be emerging.

Any response (the *R* in RTI) to intervention must assume high-quality instruction for all students. Diagnostic assessments screen for early detection of mathematics needs. Instructional actions are then implemented to support students in need (Fennell 2010). The cycle then includes follow-up assessments to determine whether students have made adequate progress and thus no longer need intervention, continue to need some level of intervention support, or need more intensive intervention. Researchers typically refer to the levels of intervention as three tiers with specific characteristics (Fuchs, Fuchs, and Vaughn 2008), although some settings include more than three tiers for the intervention support. Tiers 1–3 may be defined as follows:

◆   Tier 1—typically classroom-based intervention aligned with the specific mathematics curriculum topics that all students in the class receive. All students undergo assessment or screening, and the classroom teacher typically determines and implements particular interventions (e.g., more time on a concept, use of a particular model, more practice).

◆   Tier 2—students receiving tier 2 interventions have typically demonstrated a greater need for targeted assistance in key mathematics concepts. Such intervention may occur in small-group settings in the classroom or take the form of supplemental instruction from the classroom teacher, a mathematics specialist, or an instructional assistant. This additional time for mathematics will vary but may range from twenty to forty minutes four to five times a week (Fuchs, Fuchs, Craddock, et al. 2008).

◆   Tier 3—students needing tier 3 interventions require more intensive assistance. Such programs are likely to be supplemental and occur outside the daily mathematics lesson. One-on-one tutoring and additional supplemental support in mathematics tend to be the norm at the tier 3 level. Special education services and professionals may be involved at this level as the interventionist, although sometimes the classroom teacher may have this responsibility. At the tier 3 level foundational mathematics topics must be central to the instructional activities delivered.

Although intervention and RTI may be considered a by-product of NCLB/ESEA and IDEA, the concept has become a lifeline for students who struggle with mathematics and must be a carefully considered opportunity for students who are in need—regardless of instructional level or how schools and school districts define and address tiers (Fennell 2010). As readers consider this book's chapters related to learning, instruction, assessment, and mathematics content, thoughts regarding application to local and regional RTI initiatives may be an important consideration.

Diagnosis

One could generalize that at some point most students need some level of intervention—just to get caught up. This process may include a few quiet minutes going over solution strategies for how to share 4 candies among 6 people, giving students time to get underneath why a particular representation works for showing that first each of the 6 people would get half of a candy and then another $\frac{1}{6}$ of a candy, and discussing that although the sharing part may be easy, thinking about how much candy each person now has may be harder to both represent and, importantly, understand.

As teachers see their students learn—and teachers do get to actually see learning taking place—they can also see the stumbling blocks, challenges, and places where their students just don't quite get it. What then? Diagnosis occurs every day in every classroom. Sometimes these are the observable moments, as the fraction-sharing example demonstrated; sometimes it's a diagnostic check at the end of the lesson to assess progress, a type of formative assessment. As this book's assessment chapter notes, the interview and other versions of formative assessment yield diagnostic cues regarding misconceptions, shallow understanding, and perhaps more significant cues for intervention or even referrals. One helpful diagnostic tool is the probe (fig. 1.1), an interview instrument designed to assess conceptual understanding, procedural fluency (as appropriate), and application of the particular concept or skill in a problem-based context. This tool gives the teacher a bit more information on levels of student understanding and proficiency—and perhaps even affect.

Every mathematics teacher needs to also be a diagnostician. An everyday need exists to quickly determine the challenges, misconceptions, lack of prerequisites, and other issues that influence student achievement and, sooner than one might think, interest in mathematics. This resource will help you with strategies in the major content domains that affect pre-K–8 students in mathematics and will have you consider learning, instructional, curricular, and assessment adaptations and techniques to better serve the needs of your students.

Curriculum

All does mean all. For mathematics curriculum, this concept means that special education students, students working through any of the common tiers of intervention, and all other children in a classroom are entitled to a full and balanced curriculum that challenges them, that appropriately meets their needs, and for which they should be held accountable. This resource book examines the curricular implications of the most recent version of the National Council of Teachers of Mathematics (NCTM) Standards—the *Principles and Standards for School Mathematics* (NCTM 2000). It also considers the *Curriculum Focal*

The following is a diagnostic tool to determine areas of instructional need, using the interview format to assess conceptual understanding, computational fluency (as appropriate), problem solving, and disposition. Probes may be used as pre- and postassessments on particular areas of need or just to identify levels of understanding or proficiency. The teacher should ask a student to talk through his or her responses to the interview questions. When finished, teachers may want to make notes regarding statements made. Students should respond in pen, so that teachers can analyze elements of the response that students scratched out.

- **Conceptual understanding.** Provide one to two items designed to show whether the student understands the process(es) in the area of instruction. This activity could be manipulative based. For example, if the topic is multiplying two-digit by one-digit numbers, ask the student to shade in the following grid (using different colors, if possible) to show how one can think of 17 × 5 as (10 × 5) + (7 × 5) or as 50 + 35.

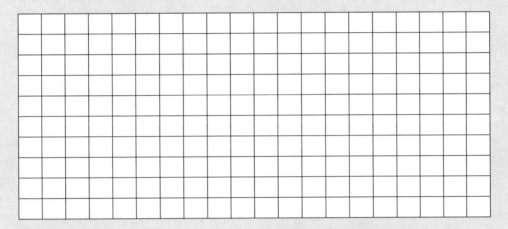

- **Computational fluency.** Supply enough examples (typically five to seven) for the student to demonstrate areas of need or show strengths. So, following the preceding example:

$$\begin{array}{ccccc} 13 & 24 & 35 & 46 & 56 \\ \underline{\times\ 8} & \underline{\times\ 9} & \underline{\times\ 4} & \underline{\times\ 6} & \underline{\times\ 5} \end{array}$$

- **Problem solving.** Offer a problem-based context for the concepts and skills being assessed. Provide one to two examples, such as the following:

  Susan drove the 28-mile round trip from Bethany Beach, Delaware, to the southernmost point of Ocean City, Maryland, each day for one full week. How far did she travel that week? Show your work and label your answer.

- **Disposition.** Consider posing one to two items about interest in relation to the topic being assessed. For example:

  What do you like about multiplication? Which was easier, the multiplication examples or the word problem?

Fig. 1.1. The probe, a diagnostic interview tool

Points for Prekindergarten through Grade 8 Mathematics (NCTM 2006). The *Common Core State Standards* (CCSSO 2010) present a third window of analysis for the curricular adaptations that must be considered essential for special education students and those students receiving varied tiers or levels of intervention.

A constant challenge for many teachers is considering how much time they need to spend on core expectations at various grade levels. The National Mathematics Advisory Panel (U.S. Department of Education 2008) identified Critical Foundations as "must haves" before students complete a formal course in algebra. The Critical Foundations include proficiency with whole numbers, fractions (defined as *a/b* fractions, decimals, and percent), and particular aspects of geometry. All students should acquire such a foundational background, but this would not be a full curriculum at the pre-K–8 levels.

A related approach would be to consider the *Curriculum Focal Points* (NCTM 2006) as the essentials for particular grade levels, with the connections for each grade level's focal points to expand beyond these minimums. The fourth-grade Curriculum Focal Points (see fig. 1.2) serve as one grade-level example for such a focus. The discussion of the mathematics under each Focal Point (or connection) includes several crucial mathematical understandings that lead to the proficiency expected in the Focal Point. Also, at the top of the Focal Point page is a direct reference to engaging students in the mathematical processes in *Principles and Standards* (NCTM 2000). Such opportunities to solve problems, reason, communicate, connect to other mathematics and/or subjects, and use representation are, all too often, neglected for far too many students.

Another consideration for curricular adaptation is to use the *Common Core State Standards* (CCSSO 2010) as the blueprint for curricular expectations. These standards include eight mathematical practices that rest on processes and proficiencies—the processes of the NCTM *Principles and Standards* (NCTM 2000) and the proficiencies from *Adding It Up: Helping Children Learn Mathematics* (Kilpatrick, Swafford, and Findell 2001). The Standards for Mathematical Practice are ways in which students should be engaged in the mathematics they are learning—regardless of grade or instructional level. The Practices should regularly intersect with the Common Core's content standards. The Standards for Mathematical Practice are as follows:

1. Make sense of problems and persevere in solving them.
2. Reason abstractly and quantitatively.
3. Construct viable arguments and critique the reasoning of others.
4. Model with mathematics.
5. Use appropriate tools strategically.
6. Attend to precision.
7. Look for and make use of structure.
8. Look for and express regularity in repeated reasoning.

Ensuring that all mathematical learning experiences engage the practices is just one way

Grade 4 Curriculum Focal Points	Connections to the Focal Points
***Number and Operations and Algebra:* Developing quick recall of multiplication facts and related division facts and fluency with whole number multiplication** Students use understandings of multiplication to develop quick recall of the basic multiplication facts and related division facts. They apply their understanding of models for multiplication (i.e., equal-sized groups, arrays, area models, equal intervals on the number line), place value, and properties of operations (in particular, the distributive property) as they develop, discuss, and use efficient, accurate, and generalizable methods to multiply multidigit whole numbers. They select appropriate methods and apply them accurately to estimate products or calculate them mentally, depending on the context and numbers involved. They develop fluency with efficient procedures, including the standard algorithm, for multiplying whole numbers, understand why the procedures work (on the basis of place value and properties of operations), and use them to solve problems.	***Algebra:*** Students continue identifying, describing, and extending numeric patterns involving all operations and nonnumeric growing or repeating patterns. Through these experiences, they develop an understanding of the use of a rule to describe a sequence of numbers or objects. ***Geometry:*** Students extend their understanding of properties of two-dimensional shapes as they find the areas of polygons. They build on their earlier work with symmetry and congruence in grade 3 to encompass transformations, including those that produce line and rotational symmetry. By using transformations to design and analyze simple tilings and tessellations, students deepen their understanding of two-dimensional space.
***Number and Operations:* Developing an understanding of decimals, including the connections between fractions and decimals** Students understand decimal notation as an extension of the base-ten system of writing whole numbers that is useful for representing more numbers, including numbers between 0 and 1, between 1 and 2, and so on. Students relate their understanding of fractions to reading and writing decimals that are greater than or less than 1, identifying equivalent decimals, comparing and ordering decimals, and estimating decimal or fractional amounts in problem solving. They connect equivalent fractions and decimals by comparing models to symbols and locating equivalent symbols on the number line.	***Measurement:*** As part of understanding two-dimensional shapes, students measure and classify angles. ***Data Analysis:*** Students continue to use tools from grade 3, solving problems by making frequency tables, bar graphs, picture graphs, and line plots. They apply their understanding of place value to develop and use stem-and-leaf plots. ***Number and Operations:*** Building on their work in grade 3, students extend their understanding of place value and ways of representing numbers to 100,000 in various contexts. They use estimation in determining the relative sizes of amounts or distances. Students develop understandings of strategies for multidigit division by using models that represent division as the inverse of multiplication, as partitioning, or as successive subtraction. By working with decimals, students extend their ability to recognize equivalent fractions. Students' earlier work in grade 3 with models of fractions and multiplication and division facts supports their understanding of techniques for generating equivalent fractions and simplifying fractions.
***Measurement:* Developing an understanding of area and determining the areas of two-dimensional shapes** Students recognize area as an attribute of two-dimensional regions. They learn that they can quantify area by finding the total number of same-sized units of area that cover the shape without gaps or overlaps. They understand that a square that is 1 unit on a side is the standard unit for measuring area. They select appropriate units, strategies (e.g., decomposing shapes), and tools for solving problems that involve estimating or measuring area. Students connect area measure to the area model that they have used to represent multiplication, and they use this connection to justify the formula for the area of a rectangle.	

Fig. 1.2. Grade 4 Curriculum Focal Points (from NCTM 2006)

of ensuring appropriate curricular opportunities for all students. Another consideration is the *Common Core State Standards* themselves. The Common Core's grade-level standards designate critical areas that "instructional time should focus on" (CCSSO 2010, p. 27) in each grade level. Consider the critical areas for grade 4: "In Grade 4, instructional time should focus on three critical areas: (1) developing understanding and fluency with multi-digit multiplication, and developing understanding of dividing to find quotients involving multidigit dividends; (2) developing an understanding of fraction equivalence, addition and subtraction of fractions with like denominators, and multiplication of fractions by whole numbers; (3) understanding that geometric figures can be analyzed and classified based on their properties, such as having parallel sides, perpendicular sides, particular angle measures, and symmetry" (CCSSO 2010, p. 27).

The critical areas do not fully encompass mathematics learning expectations for a particular grade, but they do indicate priority topics, similar to the Curriculum Focal Points (NCTM 2006). A grade-level analysis adapted from the *Common Core State Standards* displays the mathematical domains and standards for each grade. (See the following grade 4 example.) For a finer-grained review, consult all the cluster standards, with more specific expectations (see asterisk-marked items in the example) in a particular grade.

Grade 4 overview

Critical Areas

- ◆ Developing understanding and fluency with multidigit multiplication and developing understanding of division
- ◆ Developing an understanding of fraction equivalence, addition and subtraction of fractions with like denominators, and multiplication of fractions by whole numbers
- ◆ Understanding that geometric figures can be analyzed and classified on the basis of their properties

Domain: Operations and Algebraic Thinking

- ◆ Use the four operations with whole numbers to solve problems.
- ◆ Gain familiarity with factors and multiples.
- ◆ Generate and analyze patterns.

Domain: Number and Operations in Base Ten

- ◆ Generalize place value understanding for multidigit whole numbers.
- ◆ Use place value understanding and properties of operations to perform multidigit arithmetic.
  - • *Fluently add and subtract multidigit whole numbers by using the standard algorithm.

- *Multiply a whole number of up to four digits by a one-digit whole number, and multiply two two-digit numbers, using strategies based on place value and the properties of operations. Illustrate and explain the calculation by using equations, rectangular arrays, and/or area models.
- *Find whole-number quotients and remainders with up to four-digit dividends and one-digit divisors, using strategies based on place value, the properties of operations, and/or the relationship between multiplication and division. Illustrate and explain the calculation by using equations, rectangular arrays, and/or area models.

Domain: Number and Operations—Fractions

◆ Extend understanding of fraction equivalence and ordering.

◆ Build fractions from unit fractions by applying and extending previous understandings of operations on whole numbers.

◆ Understand decimal notation for fractions and compare decimal fractions.

Domain: Measurement and Data

◆ Solve problems involving measurement and conversion of measurement from a larger unit to a smaller unit.

◆ Represent and interpret data.

◆ Geometric measurement: understand concepts of angle, and measure angles.

Domain: Geometry

◆ Draw and identify lines and angles, and classify shapes by properties of their lines and angles.

Curricular decision making is integral to the planning and instruction elements of the teaching cycle of planning, implementation (teaching), and assessment. This resource book's chapters on instruction and the content chapters related to number and operations, algebra, geometry, measurement, and data analysis and probability offer instructional guidance and content-based considerations for such critical and daily decision making, particularly for developing student conceptual understanding and proficiency.

Assessment

As indicated earlier, all teachers use a variety of assessments to interpret student understandings, analyze student work, and assess student progress. Teachers need to match types and levels of assessment to what is being assessed. Assessments must be constructed and used to meet the specific needs of students. This resource book's chapter on assessment presents a far more detailed discussion of the important role of assessment in teaching and guiding the development of student understanding and proficiency. However, the following considerations related to formative and summative assessment are beginning points for this work.

Formative assessments

In-class formative assessments form the basis for daily decision making related to student progress and instructional needs. Formative assessment opportunities include observing students as they learn. Observations help teachers determine how engaged students are in a lesson's activity and their level of success with the activity; observations can also yield on-the-spot interventions that can make a particular lesson or activity more successful for students. Classroom conversations are another integral component of formative assessment. Students need opportunities to discuss their thinking. Formative assessments may also include using interviews; rubrics; academic and behavior checklists; surveys; journals and portfolios; audio, video, or smart-pen artifacts; and additional student reflections. Use of student writing and one-on-one or small-group interviews are helpful techniques for identifying student misconceptions early, thus offering opportunities for intervention. Routine classroom activities, including homework, quizzes, and projects, also afford opportunities for assessing levels of student understanding and progress toward curricular expectations.

Teachers assess as they teach. They observe, ask questions, monitor student work, and generally track the success of their lessons through formative assessment. They understand curriculum expectations and students well enough to know when a lesson is not working and when students are lost. Formative assessment can guide instruction—every day. Feedback to students on the basis of formative assessment linked with specific suggestions for intervention also improves mathematics achievement (Gersten, Baker, and Chard 2006).

Teachers know the benefits and limitations of different formative assessment methods and can justify selecting them, including those based on the unique needs of their students. When possible, teachers must match assessment techniques to students' developmental levels and to the particular concepts and skills being assessed.

Summative assessments

Summative assessments are measures with some distance from the day-to-day work of the classroom teacher. These are the unit tests, school district–mandated benchmarks, and NCLB/ESEA state assessments administered throughout the instructional year. Such measures, if used appropriately and not overemphasized, can offer guidance and direction as to student, school, and district comparisons and progress toward national, state, and local curricular expectations. Many summative assessments are standardized. Although summative assessments (particularly standardized) have limitations, they can also create opportunities for planning for instruction. Such analyses could include data discussions about instructional and curricular decisions or could help teacher colleagues determine how their school district, school, and class is doing comparatively. What targets should be considered for improving grade-level progress? What instructional strategies, materials, or practices can be put in place to assist students? From another perspective, summative assessments enable a surface-level diagnosis regarding content and perhaps pedagogy. Suppose that more than half the sixth graders on an instructional team missed a state assessment item related to proportion that involved tables of equivalent ratios. Discussing how much opportunity the students have had to solve ratio and proportion problems by using ratio tables would make sense, as well as discussing how well students understand ratios and equivalent ratios (proportions).

Teachers must use the results of all assessments, formative and summative, to identify students whose learning problems have gone unrecognized and to monitor the progress of all students. Regardless of the level or method of assessment used, teachers must focus on gauging students' conceptual understandings, critical thinking, and ability to solve problems—not just acquire skills. As a crucial component of planning and instruction, assessment measures breadth and depth of learning and identifies misconceptions that may influence students' thinking. But assessment is more than item analysis, proficiency levels, and making AYP. It's also monitoring the affective and expressive qualities of student engagement. One important role of assessment is communicating and interpreting results with and for parents and others. Finally, assessments should not be done *to* students; rather, assessments are *for* students and should be used to guide and enhance their learning (NCTM 2000).

Conclusion

All students must have access, every day, to a mathematics teaching and learning environment and related experiences that meets their needs, that challenges them, and for which they are held accountable. We must get away from the sort of impulsive reaction of thinking that children who struggle need to be considered for special education rather than taking the time to determine actual needs and forging an intervention path. We must also find ways for teachers of mathematics and special education teachers to truly and regularly collaborate in areas related to learning, instruction, and assessment, as well as in the mathematics content and related pedagogy to be considered for special education students.

As I close this chapter, allow me two opportunities for disclosure. First, references to children and teachers do not use actual names of those characterized in problem contexts and chapter vignettes. Second, the book's title, *Achieving Fluency: Special Education and Mathematics*, is designed to focus on the word *fluency* as an expectation we would like for all students. "Developing fluency requires a balance and connection between conceptual understanding and computational proficiency" (NCTM 2000, p. 35), so although fluency is typically used in reference to computation, we intend to broaden the concept to include proficiency in and with algebra, geometry, measurement, and data analysis and probability.

I hope that this book becomes your on-the-shelf resource for thinking hard about students who experience challenges in learning mathematics at the pre-K–grade 8 levels. The chapters that follow will take you through important considerations regarding learning, instruction, assessment, and the content areas for which you and your students are responsible. Although special education is a formal placement decision and we have developed this resource to assist mathematics and special education teachers, the book's intent is to remind you about the essentiality of good teaching and that mathematics is special. Mathematics is special because of its complexities. Mathematics involves a wide variety of important concepts and skills, and these concepts and skills increase in complexity over the school-age years (Mazzocco 2007). So, what's so important about special education *and* mathematics? Everything (Fennell 2007).

References

Council of Chief State School Officers (CCSSO). *Common Core State Standards for Mathematics.* Washington, D.C.: CCSSO, 2010.

Fennell, Francis (Skip). "President's Message: What's So Special about Special Education? Everything!" *NCTM News Bulletin* (October 2007): 3.

———. *FocusMath Intensive Intervention.* Glenview, Ill.: Pearson, 2010.

Fuchs, Douglas, Lynn S. Fuchs, and Sharon Vaughn, eds. *Response to Intervention: A Framework for Reading Educators.* Newark, Del.: International Reading Association, 2008.

Fuchs, Lynn S., Douglas Fuchs, Caitlin Craddock, Kurstin N. Hollenbeck, Carol L. Hamlett, and Christopher Schatschneider. "Effects of Small-Group Tutoring with and without Validated Classroom Instruction on At-Risk Students' Math Problem Solving: Are Two Tiers of Prevention Better than One?" *Journal of Educational Psychology* 100, no. 3 (2008): 491–509.

Gersten, Russell, Scott Baker, and David Chard. *Effective Instructional Practices for Students with Difficulties in Mathematics—Findings from a Research Synthesis.* Annapolis, Md.: Center on Instruction Mathematics Summit, November 13, 2006.

Individuals with Disabilities Education Improvement Act (IDEIA). Public Law 108-446, 118 Stat. 2647, 2004.

Kilpatrick, Jeremy, Jane Swafford, and Bradford Findell, eds. *Adding It Up: Helping Children Learn Mathematics.* Washington, D.C.: National Academies Press, 2001.

Mazzocco, Michèle M. M. "Defining and Differentiating Mathematical Learning Disabilities and Difficulties." In *Why Is Math So Hard for Some Children? The Nature and Origins of Mathematics Learning Difficulties and Disabilities*, edited by Daniel B. Berch and Michèle M. M. Mazzocco, pp. 36–40. Baltimore: Paul H. Brookes Publishing, 2007.

National Council of Teachers of Mathematics (NCTM). *Principles and Standards for School Mathematics.* Reston, Va.: NCTM, 2000.

———. *Curriculum Focal Points for Prekindergarten through Grade 8 Mathematics: A Quest for Coherence.* Reston, Va.: NCTM, 2006.

U.S. Department of Education. *Foundations for Success: The Final Report of the National Mathematics Advisory Panel.* Washington, D.C.: U.S. Government Printing Office, 2008.

CHAPTER 2

Learning: A Framework

Arthur J. Baroody

Aaron was not considered at risk for academic failure; he *was* academic failure. Only in second grade, he was already classified as a low achiever and would soon be labeled learning disabled. Mathematics, in particular, was a struggle. The boy had to count on his fingers to determine the sum or difference of basic (single digit) number combinations such as 5 + 3 and 5 – 3. He had quickly learned to be ashamed of his informal strategies. His teacher had caught him early in the school year counting on his fingers and exclaimed, "Aaron, stop using baby math and use your head!" Thereafter, he made sure to do his finger calculations out of sight and as quickly as possible. Unfortunately, this approach often resulted in incorrect answers and further humiliation.

When his class played the "around the world" game to practice the basic combinations, Aaron made up excuses to leave the room. When this tactic failed, as it so often did, Aaron would try to hide behind the middle portion of the line, hoping he would be overlooked. Invariably, this ploy did not work: he could not respond with a sum or difference quickly and accurately, and he was eliminated in the first round of the game. As he walked dejectedly back to his seat, he taunted those who had failed before him: "At least I wasn't the first to go down like the rest of you losers." Although told to listen and learn from the others, Aaron and most of the other losers usually escaped their embarrassment by daydreaming.

The teacher's explanation of three-digit subtraction utterly baffled Aaron. However, he had learned long ago not to ask questions such as "Why do you rename?" These questions were typically not answered and, on at least one occasion, met with the refrain, "Yours is not to question why but to do or stay after school." Like his classmates, Aaron viewed arithmetic procedures as "math magic." He, like many children, concluded that he was simply too stupid to get it.

Because he was still functioning at a kindergarten level, Aaron was scheduled for extra help with the special education teacher. Another classmate, Bryan, who was functioning at a first-grade level, accompanied him. While their class studied science and learned interesting facts about dinosaurs, tornados, stars, and other intriguing things, the exiles worked on extra math worksheets. The special education teacher believed that, if children were empty vessels, then learning-disabled children were leaky vessels who needed to be overfilled—needed to "overlearn" material by massive amounts of practice—to retain anything.

Prodded by the special education teacher, Aaron listlessly started on his worksheet. He tried to recall his classroom teacher's instruction about "renaming," but because the procedure made no sense to him, he could not remember it. He proceeded to solve the problem in the best way he could. "Always subtract the smaller from the larger," he thought to him-self—something that he perhaps heard in class. Here is his work for the first item:

$$
\begin{array}{r}
4\ \ 0\ \ 5 \\
-\ \ \ \ 7\ \ 9 \\
\hline
4\ \ 7\ \ 4
\end{array}
$$

The special education teacher asked, "Does an answer of 474 make sense?" Sensing a trap but clueless about how to escape, Aaron said nothing. The special education teacher re-explained the same math magic that his classroom teacher had uttered, but more slowly. When Aaron persisted with his patently incorrect procedure, the special education teacher took out some blocks and did some things with them. "See," she said, "this is renaming."

Aaron was thoroughly confused. He had thought that "renaming" was math magic with written numbers. Now it appeared that one could do math magic with blocks also. Even after several explanations of the renaming procedure and several demonstrations with blocks, Aaron continued to make the small-from-large error (the error made on the first item shown earlier).

Bryan, who had been quietly completing his own worksheet, observed Aaron's fruit-less efforts to use the blocks and mocked, "Aaron, stop using baby math and use your head!" The humiliation was too much; Aaron threw a block as hard as he could and hit Bryan on the ridge just above the right eye. After the special education teacher had staunched the flow of blood and comforted Bryan to the point where he was merely whimpering, she marched Aaron to the principal's office. The principal scolded Aaron and then delivered a temporary reprieve from the constant humiliation: a one-week sus-pension from school.

After his suspension, Aaron's classroom teacher gave him a pile of makeup worksheets, along with the extra practice work from his special education teacher. Even after all this extensive practice, the boy continued to be perfectly error prone.

"Perseveration," the special education teacher informed the group assembled to discuss Aaron's case. "He makes the same mistake again and again. He is inattentive and prone to associative confusions such as '4 − 1 is 5.' He has almost no memory for basic number facts. He is still reversing his numerals. These are classic symptoms of a learning disability. He needs to be tested. Furthermore, his regular acting out is indicative of a behavior disorder."

Aaron sensed that being the only one taken out for testing was not good, despite the school psychologist's assurances. "He is trying to prove how stupid I am," Aaron thought to himself. Full of despair and panic, he tried to get through the testing as quickly and with as little thought as possible. The subsequent parent–teacher conference confirmed Aaron's worst fears when the school psychologist announced, "He's learning disabled."

The special education teacher gave Aaron even more help, largely in the form of three times the practice work given his classmates. Despite the best intentions of his teachers and all their extra effort, Aaron fell further and further behind his classmates and became more and more disruptive.

The National Council of Teachers of Mathematics (NCTM 1989, 1991, 2000, 2006) advocates fostering the *mathematical proficiency* of *all* children. Fostering mathematical proficiency implies cultivating *adaptive expertise* (meaningful knowledge that can be thoughtfully and flexibly adapted and applied to new tasks), which Hatano (1988) contrasts with *routine expertise* (knowledge memorized by rote—knowledge that can be used effectively with familiar, but not unfamiliar, tasks). By "all children," NCTM (1991) includes "students who have not been successful in school and in mathematics" (p. 125), such as children with learning difficulties or behavioral disorders (e.g., Thornton and Bley 1994). "Children with learning difficulties" encompasses students with learning disabilities or mental retardation, as well as those who exhibit low mathematics achievement.

This chapter addresses three questions: (1) What exactly is "mathematical proficiency"? (2) Is it realistic to expect children with learning difficulties or behavioral disorders, such as Aaron, to achieve mathematical proficiency in any real sense? (3) How can teachers maximize what mathematical potential these children do have?

What Is Mathematical Proficiency?

Despite the seemingly endless dialogue that some refer to as the "math wars," some agreement exists about what the goals of school mathematics should be. According to *Adding It Up: Helping Children Learn Mathematics* (Kilpatrick, Swafford, and Findell 2001) and *Foundations for Success: The Final Report of the National Mathematics Advisory Panel* (U.S. Department of Education 2008), mathematics instruction should foster *mathematical proficiency*. This proficiency includes *conceptual understanding*, *procedural fluency* (e.g., fluency with basic computational skills), *strategic and adaptive mathematical thinking (problem solving and logical reasoning)*, and a *productive disposition* (the beliefs and confidence necessary to use mathematics effectively in everyday life, even as it changes constantly and rapidly; cf. Schoenfeld 1985, 1992). *Adding It Up* further recommends that instruction foster these components of mathematical proficiency in an *intertwined manner*. The following sections define and give the rationale for the four aspects of mathematical proficiency: conceptual understanding, procedural fluency, strategic and adaptive mathematical thinking, and productive disposition.

I note why each is important for other aspects of proficiency, success in school mathematics, and everyday life, as well as each aspect's key instructional implications. (For a discussion of mathematical proficiency and its components, as *Adding It Up* defines them, see Kilpatrick, Swafford, and Findell 2001).

Conceptual understanding

Conceptual understanding is knowledge of facts, generalizations, or principles underlying the comprehension of concepts (categories), relations (between categories), or operations (actions or events involving categories [Baroody, Feil, and Johnson 2007]). Superficial or weak understanding is characterized by sparse connections—few or no connections with other aspects of knowledge. Deep or strong understanding entails a dense web of connections—many interconnections with many other aspects of knowledge (e.g., Ginsburg 1977; Hiebert and Carpenter 1992). An understanding of fractions, for instance, entails recognizing that a fraction such as $\frac{2}{3}$ can be linked to various meanings such as part–whole situations (two of three equal parts of a whole), division (two units of a measure divided evenly among three groups), and ratios (two of a quantity compared to three of another quantity). As the interconnections in a web increase, understanding deepens. For example, a deep understanding of fractions includes recognizing such connections as the following: (1) Fractions, decimals, and percent are alternative representations of the same quantity (e.g., $\frac{2}{3} = 0.666\ldots = 66.6\%$). (2) One can convert a fraction into a decimal by dividing its numerator by its denominator (e.g., $\frac{2}{3} = 2 \div 3 = 0.666\ldots$), and one can convert a decimal to a percent by multiplying by 100% (e.g., $0.666\ldots \times 100\% = 66.6\%$).

Learners build up connections by two processes: *assimilation*, which involves relating new information to existing knowledge, and *integration*, which entails linking two previously separate pieces of existing knowledge. As the following description of these two processes suggests, meaningful learning often involves resolving cognitive conflict and gradually moving from a relatively incomplete and inaccurate understanding to a relatively complete and accurate one. As the descriptions further suggest, meaningful learning is not merely a process of adding information but also one that transforms thinking (DeRuiter and Wansart 1982).

- ◆ *Assimilation.* Learners filter, comprehend, and remember new information in light of what they already know. Interpreting and incorporating new information in the context of existing knowledge is called assimilation (e.g., Piaget 1964). During assimilation, a child may encounter aspects of the new experiences that do not exactly fit existing knowledge. This experience creates conflict or disequilibrium, which requires an adjustment. Adjusting existing concepts or strategies to meet the demands of a new experience is called *accommodation.* Through the complementary processes of assimilation and accommodation, then, knowledge becomes enriched and more responsive to the environment. That is, we construct a more complete and accurate understanding of the world and better-adapted ways of responding to it.

- ◆ *Integration.* By reflecting on experiences, a child may recognize that two previously disconnected aspects of existing knowledge are, in fact, related. This scenario is especially likely when an experience causes two aspects of knowledge to come into conflict (e.g., fig. 2.1). The resulting disequilibrium causes a mental reorganization: combining and refining the two aspects of knowledge. The result is a broader perspective, one that is more complete and accurate.

Learning school mathematics in a meaningful, well-connected manner creates the basis for adaptive expertise. Indeed, such learning is a key basis for achieving all four aspects of mathematical proficiency. Well-connected instruction makes attaining conceptual understanding, procedural fluency, strategic and adaptive mathematical thinking, and a productive disposition more likely than learning by rote for five reasons:

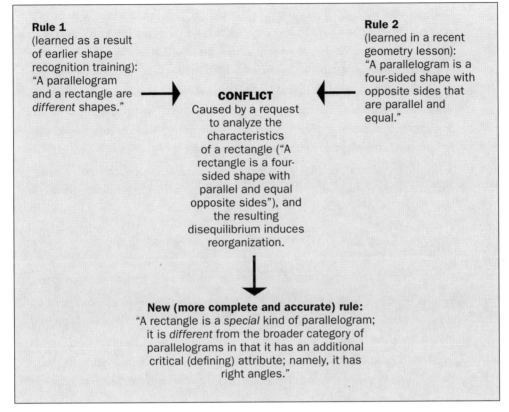

Rule 1
(learned as a result of earlier shape recognition training): "A parallelogram and a rectangle are *different* shapes."

CONFLICT
Caused by a request to analyze the characteristics of a rectangle ("A rectangle is a four-sided shape with parallel and equal opposite sides"), and the resulting disequilibrium induces reorganization.

Rule 2
(learned in a recent geometry lesson): "A parallelogram is a four-sided shape with opposite sides that are parallel and equal."

New (more complete and accurate) rule:
"A rectangle is a *special* kind of parallelogram; it is *different* from the broader category of parallelograms in that it has an additional critical (defining) attribute; namely, it has right angles."

Fig. 2.1. Integration of conflicting geometry ideas

1. Students' initial learning of facts, procedures, and formulas (achieving procedural fluency), as well as definitions and principles (mathematical concepts), is more likely and easier with well-connected instruction. When children see connections between, say, a renaming procedure and grouping/place-value concepts, they can learn the former more quickly than by memorizing it by rote, which typically requires considerable drill and practice. For example, if children understand that 304 represents three hundreds, no tens, and four ones, they are more likely to comprehend that to subtract 8 from it, one of the three hundreds must be traded for ten tens, one of which can be traded for ten ones, yielding ten ones and four ones, or fourteen ones.

2. Students are more likely to remember understandable facts, procedures, and formulas than incomprehensible knowledge learned by rote. This retention further reduces the amount of review and further practice needed to master content knowledge (to achieve procedural fluency).

3. Students can apply well-connected knowledge to familiar situations flexibly and thus increase procedural fluency. For instance, understanding how fractions and decimals are related can help students choose the easiest procedure for multiplying two rational numbers. For example, determining the product of one-third of three-fourths is easier to do with fractions ($\frac{1}{3} \times \frac{3}{4} = \frac{1}{4}$) than with the decimal equivalents ($0.333\ldots$ of $0.75 = 0.2499999975$). In contrast, determining the product of 10% of $24 is easier to do with decimals ($0.10 \times \$24 = \$2.40$) than with fractional equivalents ($\frac{1}{10} \times \$24 = \$12/5 = \$2.4$).

4. Students can more readily, flexibly, and effectively apply conceptually informed, understood knowledge to (somewhat) new situations. This approach can facilitate learning new material or solving new and challenging problems (offering a basis for strategic and adaptive mathematical thinking).

5. Students are less intimidated by material that makes sense and, thus, are more likely to enjoy and welcome learning and using mathematics (to develop a productive disposition).

▼ **Key Instructional Implications**

The goal of conceptual understanding implies the following:

A major focus of instruction should be on *helping students see how new material is connected to what they already know, including their often rich everyday (informal) knowledge of mathematics.*

Assimilation is most likely to engage interest and yet be sufficiently meaningful to students if *new material (e.g., a problem or idea) is just beyond what they already know (i.e., is moderately novel).*

A major focus of instruction should be on *helping students relate different aspects of their knowledge.*

Teachers can spur this process of integration by *creating situations where different views come into conflict.*

Procedural fluency

Procedural fluency involves using basic skills such as facts, procedures, and formulas efficiently (i.e., quickly and accurately). It also entails knowing when to use them and, if necessary, how to adapt them. In other words, procedural fluency is skill in carrying out routines *appropriately* and *flexibly* as well as *efficiently* (Kilpatrick, Swafford, and Findell 2001; NCTM 2000).

General agreement exists that students need to be able to use basic skills efficiently so that they can focus on more complex issues or efforts. However, skill mastery must involve adaptive expertise rather than mere routine expertise. Put differently: children must learn facts, procedures, algorithms, and formulas meaningfully to apply such basic skills appropriately and, if necessary, flexibly modify or adapt their application to new tasks or problems.

▼ **Key Instructional Implications**

The goal of procedural fluency implies the following:

Mastery of basic skills is essential.

This mastery must be tied to conceptual understanding.

Strategic and adaptive mathematical thinking

Kilpatrick, Swafford, and Findell (2001) distinguished between strategic competence and adaptive reasoning. The former involves the ability to formulate, represent, and solve mathematical problems, and the latter entails the capacity for logical thought, reflection, explanation, and justification. (I will not distinguish between these two competencies but will use the more general term *strategic and adaptive mathematical thinking* for two reasons: [1] The two components are intricately interrelated. For example, logical reasoning is a key strategy for solving problems. [2] Mathematical thinking involves an exploratory side as well as a logical side [Clements and Battista 1992].) *Adding It Up* (Kilpatrick, Swafford, and Findell 2001) correctly emphasizes the logical side (the importance of deductive reasoning) but makes little mention of the exploratory side of mathematics, namely, intuitive and inductive reasoning. Intuitive and inductive reasoning serve as the bases for formulating conjectures and informally evaluating conjectures by, for instance, looking for examples and nonexamples. In brief, strategic and adaptive mathematical thinking encompasses all the processes involved in mathematical inquiry.

To achieve mathematical proficiency, students need more than knowledge of mathematical content (an accumulation of facts, definitions, procedures, formulas, and rules). They need to know strategies for how to think about and apply this knowledge. Put differently: they need the acumen to engage in mathematical inquiry (e.g., representing, problem solving, reasoning, and communicating). See the NCTM (2000) Standards for a discussion of these processes of mathematical inquiry.

▼ **Key Instructional Implications**

The goal of strategic and adaptive mathematical thinking implies the following:

Students need regular, supervised practice in solving mathematical problems and engaging in the other processes of mathematical inquiry.

To encourage autonomy, teachers should prompt students to devise their own methods for solving problems and to evaluate their own methods and solutions to the extent possible.

Productive disposition

A productive disposition entails believing that mathematics makes sense and is useful, that learning it requires diligence, and that everyone is capable of significant mathematical learning. It involves viewing mathematics as understandable and personally worthwhile, appreciating that learning or applying mathematics requires persistence and discipline, and having the self-confidence to take on the challenges of learning or applying mathematics. The National Mathematics Advisory Panel (U.S. Department of Education 2008) indicates that children's beliefs about the relative importance of effort and ability or inherent talent can be changed and that increased emphasis on the importance of effort is important. A productive disposition relative to mathematics can and should be nurtured.

When children view mathematics as sensible and personally worthwhile, they are naturally more motivated to learn it and stick to a problem or task. Because they can apply conceptual understanding to a task, they are more likely to devise or use an appropriate strategy and arrive at a successful solution. Such success rewards their determination, making them more likely to become persistent learners or problem solvers. Real success also fosters real self-confidence.

▼ **Key Instructional Implications**

The goal of a productive disposition implies that instruction should (1) be meaningful and purposeful, (2) require discipline and perseverance, and (3) foster productive beliefs (e.g., everyone is capable of significant mathematical learning).

* * *

We have to ask ourselves whether mathematical proficiency is a realistic goal for children with learning difficulties. In the following section, I first discuss why the goal of mathematics instruction for typically developing children has shifted from routine expertise to adaptive expertise—from memorizing basic skills by rote to achieving all aspects of mathematical proficiency. Next, I explain why the special education community has not embraced this same shift more widely. Finally, I address why promoting mathematical proficiency makes sense for all children, including those with learning difficulties and behavioral disorders.

A New Direction in Mathematics Education

Scholars and teachers have debated how mathematics should be taught since the time of the ancient Greeks (Wilson 2003). In the early twentieth century, Edward L. Thorndike (1922), who helped lay the foundations for a behavioristic psychology, summarized his research on direct instruction and drill in *The Psychology of Arithmetic*. His main point was that curricula and practice had to be organized carefully so that children developed habits (memorized skills by rote) efficiently. Reformers such as William Brownell (1935), who built on what would later be called cognitive psychology, pointed out the limitations of such an approach and advocated meaningful instruction. Although Brownell won the hearts and minds of mathematics education researchers, Thorndike won the hearts and minds of practitioners of mathematics education (Ginsburg, Klein, and Starkey 1998). Historically, then, mathematics instruction in the United States has typically been based on the model of direct instruction and drill that Thorndike advocated. For a variety of reasons, NCTM (1989, 1991, 2000, 2006) has recommended a radically different approach to instruction, one more attuned to Brownell's views than those of Thorndike.

A traditional direct-instruction-and-drill approach

Here I outline the basic assumptions of traditional instruction, its nature, and its impact on promoting mathematical proficiency.

Basic assumptions

The direct-instruction-and-drill approach (the methods that Aaron's classroom teacher used) is based on the following assumptions:

- ◆ The mathematical knowledge that nearly all people need are basic skills, which include socially useful facts (number, arithmetic, geometry, or measurement), definitions, rules, procedures, and formulas.
- ◆ Children come to school with little or none of this content knowledge, are largely incapable of strategic or adaptive mathematical thinking, and are not inherently inclined to learn or apply mathematics.
- ◆ The most efficient way to help students master basic skills is to tell or show them what they need to know and have them practice the skills until mastery. Conceptual understanding and strategic and adaptive mathematical thinking are not necessary to accomplish this goal.
- ◆ Teachers can overcome children's natural lack of interest in learning school mathematics with the promise of rewards (e.g., a sticker, candy bar, a smile, extra play time, or a good grade) or the threat of punishment (e.g., a frown, extra work, reduced play time, or a failing grade).

The nature of instruction

As Aaron's situation illustrated, traditional instruction is usually highly abstract and based on memorizing basic skills by rote. More specifically, it typically consists of verbal directions or rules regarding written symbols or procedures. Furthermore, this instruction is almost always done in a vacuum or without context—unrelated to real, meaningful, and purposeful activities or experiences. Students memorize knowledge by frequent and extensive practice, often in flashcard drills and row upon row of assigned textbook or worksheet problems.

Impact on mathematical proficiency

Direct-instruction-and-drill approaches to mathematics—even those that are well designed—fail to foster the four aspects of mathematical proficiency that *Adding it Up* (Kilpatrick, Swafford, and Findell 2001) defines, including procedural fluency.

◆ With the basic assumptions and nature of the traditional approach, it is not surprising that children in such programs typically have little or no understanding of the material they learn. Because procedures (how-to knowledge) are not connected to concepts (knowledge of why), other aspects of mathematics, or everyday life, they are often meaningless to students.

◆ Although many children master basic skills by rote and use them efficiently on familiar tasks, even these "successful students" may not recognize appropriate applications in everyday life or other contexts. Moreover, students often have difficulty flexibly adapting meaningless skills when they encounter new tasks, including those that are only somewhat unfamiliar (e.g., Wertheimer 1959). The typical result is incomplete procedural fluency.

◆ Because students are simply told how to solve particular types of problems and are then expected to imitate these procedures, instruction is not inquiry based. Not surprisingly, students seldom develop the strategic and adaptive mathematical thinking necessary to solve challenging (or even simple) problems.

◆ Since traditional instruction usually has little or no purpose or relation to children's everyday life or interests (other than collecting rewards and avoiding punishment), many have little interest in understanding, exploring, and applying mathematics. Other ways in which more traditional instruction undermines a productive disposition include creating the expectation that mathematical tasks should always be solved quickly and in a particular way (e.g., informal methods are inappropriate and stupid).

The current reform movement

NCTM (1989, 1991, 2000, 2006) has recommended changes in what mathematics is taught and emphasized, but perhaps more important, in how it is taught. Here I discuss the rationale for this approach, its nature, and its potential impact on children's mathematical proficiency.

Rationale for reforming mathematics instruction

Public service groups such as NCTM, state and federal governments, and business organizations became interested in reforming how and what mathematics is taught for several reasons:

◆ Our information- and technology-based society is both complex and rapidly changing. Routine expertise of basic content knowledge is no longer adequate. Adaptive expertise is essential for making the complex decisions and solving the novel problems that government, business, medicine, and so forth, confront constantly (e.g., Carnegie Forum on Education and the Economy 1986; National Commission on Excellence in Education 1983; National Commission on Mathematics and Science Teaching for the 21st Century 2000). Such expertise grows from instruction that fosters all four aspects of mathematical proficiency in an intertwined manner.

◆ Children develop significant (informal) mathematical knowledge before entering or outside of school, are capable of impressive strategic and adaptive mathematical thinking, and are inherently interested in mathematical aspects of their environment (Baroody 1987a, 1998; Clements and Sarama 2004; Court 1920; Ginsburg 1977; Ginsburg, Klein, and Starkey 1998; Kilpatrick, Swafford, and Findell 2001).

◆ Instruction that focuses on fostering learning by rote (routine expertise) is not effective in promoting mathematical proficiency, whereas instruction that focuses on promoting meaningful learning (adaptive expertise) is more likely to do so (e.g., Baroody 1998; Baroody and Dowker 2003; Davis 1984).

The nature of the investigative approach

NCTM (1989, 1991, 2000, 2006) has recommended that mathematics instruction be meaningful, inquiry based, and purposeful—the "investigative approach" (Baroody 1998). This approach embodies key aspects of developmentally appropriate practices (Bredekamp 1993) and contrasts sharply with the traditional direct-instruction-and-drill approach (table 2.1).

◆ *Meaningful instruction.* Teachers help children actively construct an understanding of mathematics by carefully building on what they already understand, particularly their informal knowledge, and by prompting reflection of what they know, including how apparently different views or different aspects of knowledge are related (i.e., promoting cognitive conflict and the assimilation or integration of knowledge).

◆ *Inquiry-based instruction.* Teachers regularly create opportunities for mathematical inquiry, including problem solving (solving challenging problems), inductive reasoning (discovering, e.g., patterns, relations, procedures, and formulas) and deductive reasoning (using logical reasoning to solve problems or prove conjectures), representing (e.g., using manipulatives or drawings to represent and solve a problem), and communicating (e.g., justifying procedures or solutions). Teachers also create a social environment that encourages questioning, inquiry, and reflection. They achieve this atmosphere, in part, by encouraging students to work in groups, share ideas, and arrive at their own solutions and conclusions; students then refine solutions and conclusions through discussions and questioning by teachers and other students.

◆ *Purposeful instruction and practice.* Teachers create a real need for exploring, learning, and applying mathematics by teaching the topic in context. One way of doing so is giving students rich or worthwhile tasks. Using games, science experiments, projects, and developmentally appropriate problems raises questions and issues, promotes exploration, and gives students practice with calculation procedures or problem-solving strategies. Integrating instruction across mathematical topics and with other content areas also achieves teaching in context.

Impact on mathematical proficiency

The investigative approach can foster all four aspects of mathematical proficiency in an intertwined manner:

◆ *Conceptual understanding.* Because one focus of the investigative approach is encouraging meaningful learning through developmentally appropriate instruction, it can promote conceptual understanding, which is the basis of adaptive expertise and all aspects of mathematical proficiency.

◆ *Procedural fluency.* Because mastery of "basic skills" is done in a meaningful and inquiry-based manner, children understand, for instance, the rationale for each step in a procedure and can justify what they are doing. This adaptive expertise better enables them to use skills efficiently, appropriately, and flexibly.

◆ *Strategic and adaptive mathematical thinking.* Conceptual understanding facilitates students' ability to engage in mathematical problem solving and reasoning. Regularly engaging in mathematical inquiry further facilitates strategic and adaptive mathematical thinking. Indeed, students cannot acquire such thinking through direct instruction; they secure it only through actually engaging in mathematical problem solving, reasoning, representing, and communicating (e.g., Resnick 1989). Therefore, mathematics instruction must be based, at least in part, on solving mathematical problems in context; involve discovery learning (which entails inductive reasoning) and testing, or applying discoveries (which can require deductive reasoning); require working in small groups (which necessitates communicating mathematical ideas clearly and convincingly); and facilitate inventing and sharing arithmetic procedures (which involves all the preceding skills).

◆ *Productive disposition.* Meaningful, inquiry-based, and purposeful instruction and practice can make mathematics interesting and challenging and, thus, motivate children to learn and use it and better appreciate its value. Such instruction can also breed confidence and constructive beliefs (e.g., the conviction that one can understand important mathematical ideas or solve challenging problems). Such affect is the product of actual achievements.

Table 2.1
Two instructional approaches

Aspect of instruction	Traditional direct instruction and drill	Investigative approach
View of knowledge	There is *one* correct procedure.	There is a choice of possible procedures and often several are highly effective and efficient.
Teaching style/view of authority	Completely authoritarian: As the expert, the teacher is *the* judge of correctness. Procedures or answers that differ from those advocated by the teacher are wrong and not tolerated. Teacher provides definitive feedback (e.g., praise for the correct answer).	Semidemocratic: Teacher or students remain committed to a method or viewpoint as long as it is effective. Teacher responds to incorrect procedures or answers by posing a question, problem, or task that prompts student reflection.
Aim of instruction	Foster routine expertise: memorization of basic skills by rote.	Foster adaptive expertise (meaningful memorization) and mathematical thinking.
Effect on mathematical proficiency	Promotes limited procedural fluency.	Promotes all aspects of mathematical proficiency.
Methods	• Teacher lectures and demonstrates. • Textbook based and largely symbolic. • Children work in isolation. • Little or no use of manipulatives or technology. • Students required to follow directions and imitate "the correct" (conventional) procedure. • Practice with an emphasis on written, sterile worksheets. Quantity of practice is key.	• Various methods are used with an emphasis on students' exploring, conjecturing about, and debating ideas (e.g., semiguided discovery learning). • Worthwhile tasks (e.g., projects, problems, everyday situations, science experiments, children's literature, math games) create a need for learning and practicing math. Textbooks serve a supporting role (e.g., a source of worthwhile tasks and resolving disagreements over definitions). • Children often work together in groups. • Use of manipulatives and technology is central to many learning tasks. • Students are encouraged to invent, share, and streamline their own concrete models and, later, written procedures (including the conventional one or equally or more efficient nonconventional ones). • Practice done purposefully. *Quality* of practice is key.

The Tradition of Direct Instruction and Drill in Special Education

Although early-childhood educators, and to a lesser extent elementary school educators, have in recent years increasingly embraced a meaningful and inquiry-based approach more consistent with Brownell's (1935) views and the NCTM (1989, 2000) Standards, most special education teachers have continued to embrace a direct-instruction-and-drill approach more consistent with Thorndike's (1922) views. Here I discuss the four interrelated reasons for this stand.

Low expectations for children with learning difficulties or behavioral disorders

Why do most children with mathematical learning difficulties fail to master basic skills (obtain routine expertise), let alone achieve mathematical proficiency (including adaptive expertise)? All too often, the learning difficulties of children such as Aaron are attributed largely or solely to *their* cognitive limitations.

Historically, learning disabilities have been attributed to a brain dysfunction, often too subtle to be detected, an organic impairment that interferes with or prevents learning of a particular kind (DeRuiter and Wansart 1982; Geary 1996; Kaliski 1962; Rourke and Strang 1983; Strauss and Kephart 1955). Although some children with learning difficulties certainly have an organic impairment (see Delazer 2003 for a review of the effects of brain injuries on mathematical competencies) and exhibit hard signs of neurological dysfunction, such as abnormal EEG (brain wave pattern) or MRI (brain activation pattern) readings, most children identified as learning disabled do not (Baroody 1996b; Coles 1978). Indeed, no convincing link to learning disabilities exists for soft signs of minimal brain damage such as fine or gross motor-coordination deficiencies, mixed or confused laterality (e.g., Aaron's numeral reversals), short attention span (e.g., Aaron's lack of concentration), or other developmental disabilities.

Today, a *learning disability* is defined as a significant developmental delay (achievement two years behind agemates) due to a cognitive deficit that interferes with students' ability to learn concepts or procedures (Berch and Mazzocco 2007). Such delays due to the lack of learning opportunities (i.e., inadequate informal or formal experiences) do not technically indicate a learning disability. However, in popular use, the term *learning disabilities* refers to any serious delay, even when the evidence for a cognitive deficit is weak, dubious, or entirely absent. As Aaron's situation illustrates, educators, parents, and children often assume that a significant learning difficulty is due to a fault within the child's mind, not to external factors such as inadequate or inappropriate instruction (Baroody, Bajwa, and Eiland 2009). Educators must distinguish between children such as Adam, who simply have lacked the opportunity to learn, and the relatively few children with genuine cognitive deficits. Such children may exhibit different symptoms and may have different learning capacities. For example, the timed arithmetic performance of children with learning difficulties varies as a function of how such difficulties are defined (Mazzocco, Devlin, and McKenney 2008).

Moreover, low expectations may prevent children with genuine cognitive difficulties from realizing their sometimes surprising mathematical potential. It has long been assumed that the mental deficiencies of children with a global developmental delay (IQ of less than 85) or a learning disability render them incapable of understanding (more advanced) mathematical concepts, memorizing basic facts or procedures, or achieving strategic and adaptive mathematical thinking (e.g., Geary 1996). Such children have often been stereotyped as "passive learners," capable only of routine expertise (Baroody 1999b; Brown et al. 1983; Cherkes-Julkowski and Gertner 1989; Ferretti and Cavalier 1991; Scheid 1990).

Children with behavioral disorders are often considered to have so little emotional control that they cannot attend long to academics (sometimes characterized as "attention-deficit syndrome") and are incapable of self-directed learning (e.g., would respond to any freedom whatsoever by going wild). The conventional wisdom is that with such a limitation, the best that can be hoped for are "instructional bandages" that do not require perseverance or understanding, such as learning how to use a calculator to do computations. It also generally assumes that the instruction of children with behavioral disorders needs to be tightly controlled.

Traditionally, then, children in general are viewed as vessels that need to be filled up with socially necessary information, and those with learning or behavioral difficulties are viewed as "broken vessels" who cannot properly take in or store needed knowledge. In effect, such children are often blamed for their own learning difficulties. Indeed, proponents of direct instruction and drill still argue that the student-centered, conceptually grounded, and inquiry-based approach that NCTM (1989) recommends may be inappropriate and ineffective for children with learning or behavior difficulties, because such children presumably need highly structured learning environments (cf. Atwater et al. 1994; Carta 1995; Carta et al. 1991). Children, particularly those with behavioral difficulties, need social structure or discipline. Discipline, in general, involves social teaching (socialization) that promotes individual responsibility and mutual respect and produces immediate and firm but fair consequences for their violation. Indeed, small-group cooperative learning, whole-class discussions or debates, and other aspects of the investigative approach cannot work effectively or at all without individual responsibility (e.g., every group member contributing to a problem-solving effort, project, or assignment) or mutual respect (e.g., disagreeing in socially acceptable ways). In fact, using the investigative approach (e.g., small-group problem-solving or learning experiences) can create real situations for learning and practicing social skills.

The relative unimportance of mathematical proficiency

Low expectations have led to restricted educational goals for many children with learning difficulties. In the past, instruction for children with a moderate global developmental delay (IQ between 25 and 50) has focused on teaching everyday survival or self-sufficiency skills such as dressing oneself or mastering a menial job. Academics, particularly mathematics, were not considered possible or appropriate for lower-functioning students. Even for children with learning or behavioral difficulties who have near-normal or normal

IQs, conventional wisdom emphasized mastering reading or self-control skills more than mathematics.

The assumption for this focus was that these children were capable of and needed only basic mathematical skills: basic number facts, arithmetic computational procedures, and other simple skills best learned by rote. Low expectations and undervalued mathematical proficiency contribute significantly to the next two causes.

Inadequate research and development

Although researchers have extensively studied learning difficulties with reading, relatively little research has addressed mathematical difficulties, particularly in conceptual learning and strategic and adaptive mathematical thinking (Delazer 2003; Jordan, Hanich, and Uberti 2003). Furthermore, inadequate attention has been given to developing effective instructional methods for implementing standard-based reforms with children having learning difficulties (Giordano 1993; Hofmeister 1993; Mallory 1994; Rivera 1993).

Inadequate training in mathematics education for special education teachers

Although mathematical disabilities may be more widespread than reading disabilities (Jordan, Hanich, and Uberti 2003), special education teachers and aides receive relatively little or no training in mathematics education or related areas, such as the psychology of mathematics learning. For example, at the University of Illinois at Urbana-Champaign, undergraduate preservice elementary-level teachers must take two mathematics content courses and two mathematics education courses. Until 2001 (when the state standards changed), the special education program required no such courses. Today, the certification requirements are relatively rigorous for a special education program but still less stringent than those for elementary education majors. Specifically, special education majors take two mathematics content courses and one mathematics methods course—one fewer than elementary education majors. Teachers without adequate preparation in mathematics or its teaching and learning tend to gravitate toward a direct-instruction-and-drill approach because, unlike the investigative approach, implementing it requires little content, pedagogical, or psychological knowledge. A teacher need only follow the directions in the teacher's edition of a textbook and assign and grade exercises. As the case of Aaron illustrates, a direct result is the largely ineffective or even destructive mathematics instruction of children with learning difficulties.

A Case for Fostering the Mathematical Proficiency of All Children

The major source of most mathematical learning difficulties is how children are taught (psychologically inappropriate instruction), not their mental equipment (organic or cognitive dysfunction).

Instructional contributions or causes of learning or behavioral difficulties

Most low achievers in mathematics are probably (to paraphrase David Elkind) "instructionally disabled," not cognitively or "learning disabled" (Baroody 1996a; Baroody and Ginsburg 1991). Moreover, ineffective instruction, namely, overemphasizing direct instruction and drill, seriously compounds the learning difficulties of children with real organic dysfunctions (e.g., children with brain injuries or mental retardation). The chronic failure and tedium that ineffective and uninspiring instruction produces undoubtedly also contributes to the acting out of those with behavior difficulties. In brief, nearly all children with learning or behavior difficulties are not achieving what potential they have—largely because of inadequate instruction.

As Aaron's story illustrates, instruction and remedial efforts for children with learning difficulties are too often based largely or wholly on a direct-instruction-and-drill approach. Remedial efforts often consist of repeatedly reviewing unlearned information, fostering "overlearning" through huge amounts of practice, and profusely rewarding progress (Moyer and Moyer 1985). Like the instruction that fostered the learning difficulties in the first place, remedial instruction is often too abstract and fails to actively engage a child's thinking and interest (van Erp and Heshusius 1986). Too often, then, a "tell, show, and drill" approach to remediation does not work and may even make matters worse (Moyer and Moyer 1985). Sometimes, efforts are made to specially adapt or adjust techniques for fostering information absorption. Although introducing special tricks or bandages to help students master facts or procedures (e.g., mnemonics or "low-stress algorithms") may be helpful in the short term, they do little to enhance mathematical understanding and proficiency in the long term (Hendrickson 1983).

Unfortunately, overemphasizing a traditional direct-instruction-and-drill approach makes learning school mathematics unduly difficult and unpleasant and robs children of mathematical proficiency (Baroody 1996a, 1998). Here is how.

Failure to promote conceptual understanding

When mathematics instruction is incomprehensible, some children—particularly those with learning difficulties or behavioral disorders, such as Aaron—have difficulty constructing a conceptual understanding of mathematics. More specifically: a traditional direct-instruction-and-drill approach does not adequately help students make connections and produces a variety of conceptual gaps in student knowledge:

◆ Gaps between symbolic school instruction and children's informal, everyday knowledge

◆ Gaps between procedures (how-to knowledge) and concepts (an understanding of why)

◆ Gaps among representations of concepts or procedures (e.g., symbolic expressions, real-world examples, and concrete manipulative models)

- Gaps among mathematical topics (e.g., not making the link that $\frac{1}{4} = 1 \div 4 = 0.25 = 25\%$)

- Gaps between mathematics and other curricular areas (e.g., graphing data is a way of organizing scientific data and finding scientific relations)

These conceptual gaps force children to memorize mathematics by rote, often producing the following difficulties:

- *Incomplete or inaccurate memorization of information.* In extreme cases, students do not even bother to memorize meaningless information. Other children memorize the information partially or inaccurately. For instance, in example A, a child could remember only part of the school-taught borrowing procedure. That is, he remembered to borrow but not to reduce. Example B illustrates another common error. The child remembered her teacher's admonition to "always subtract the smaller from the larger" but misapplies this rule to multidigit subtraction.

Example A	Example B
$13\overset{1}{\cancel{4}}$	134
$-\ 48$	$-\ 48$
196	114

- *Lack of retention.* Many students dutifully memorize prescribed material and may even do well on their tests. Unfortunately, soon after the test most students quickly forget most of what they learned, which necessitates constant review. Students spend much of second grade reviewing what they learned in first grade. They spend an even larger portion of third grade reviewing what they learned in second and first grade. And so the process continues until seventh and eighth grade, where students spend nearly the whole year reviewing previous material.

- *Lack of transfer.* Because children do not understand what they have memorized, they cannot really use it to assimilate (learn) new, even closely related, material. For example, after memorizing the borrowing procedure for two-digit expressions such as 84 – 28, children then have to be spoon-fed the borrowing procedure for three-digit expressions such as 384 – 128. This approach, in turn, does not guarantee that they will know what to do in a slightly different case such as 304 – 128, where borrowing from the tens place is not possible. Recall that Aaron, who did not understand the rationale for the borrowing procedure, responded to this item by inventing his own incorrect procedure: "borrowing over zero" (i.e., in effect renaming 304 as two hundreds and fourteen ones instead of as two hundreds, nine tens, and fourteen ones).

Failure to promote procedural fluency

As the previous subsection suggested and as the case of Aaron shows, fostering skill mastery by rote is not a particularly effective means of promoting procedural fluency. Even if students do achieve routine expertise, they may not be able to apply their knowledge to appropriate real-world settings. For example, a child may be able to complete an addition worksheet but not recognize that addition is required in a real-life situation such as item cost plus tax. Students are even less likely to apply such knowledge when solving novel problems that require them to adapt or adjust their knowledge. In brief, overemphasizing a traditional direct-instruction-and-drill approach may make achieving efficient and appropriate use of basic skills hard for children, particularly those with learning difficulties (whatever the cause). It essentially guarantees that they will not use them flexibly.

Failure to promote strategic and adaptive mathematical thinking

Typically, children with learning or behavior difficulties, such as Aaron, do not receive opportunities to engage in mathematical inquiry. Without real opportunities to represent, reason about, discuss, and solve problems, strategic and adaptive mathematical thinking is not likely to develop (Thornton and Bley 1982). The traditional direct-instruction-and-drill approach does not adequately encourage, and actually stifles, the development of strategic and adaptive mathematical thinking. For example, instead of helping children learn effective problem-solving strategies, traditional instruction has usually involved teaching superficial tricks such as the "keyword approach." This method involves looking for a word that signals a particular operation (e.g., "Whenever you see the word *left*, it means subtract"). Although a keyword approach may sometimes work, it can also lead a child astray, particularly when the child deals with nonroutine or real-world problems. For example, blindly using a *left-means-subtract* rule will not work with the following problem: "When Mrs. Jones's class went out for recess, the boys left five jackets and the girls left three. How many jackets were left?" (Kilpatrick 1985).

Failure to promote a productive disposition

When children view mathematics as mysterious and personally irrelevant (as Aaron did), they are less likely to value it, put in the effort required to learn it, or develop the confidence that they can master it or adapt it for solving new problems. For example, Aaron gave up trying to understand or even learn school mathematics because he (incorrectly) concluded that he was too stupid to do so. Overemphasizing a traditional direct-instruction-and-drill approach often creates a negative disposition toward mathematics by undermining children's interest in mathematics, their autonomy, and their beliefs about the real nature of mathematics and mathematical learning.

- ◆ *Loss of interest.* Many children quickly lose interest in incomprehensible information and tune it out. Some feel frustrated when facing a torrent of meaningless words and written symbols. Heavy doses of practice with pointless exercises can further deaden interest.

- ◆ *Helplessness.* A traditional direct-instruction-and-drill approach can breed helplessness. Because it focuses on memorizing information by rote, children do not have the conceptual understanding to learn new material or solve problems on their own (Allardice and Ginsburg 1983; Davis 1984). Because of a focus on memorizing *the* correct written procedure, a traditional direct-instruction approach discourages children from inventing or using their own informal strategies. In effect, a steady diet of meaningless instruction and drill fosters blind rule following (e.g., Holt 1964). With their routine expertise having left them chronically unprepared to cope with new tasks, children grow increasingly dependent on the teacher for solutions to even modestly new tasks or problems and for judging the correctness of their answers. In brief, spoon-feeding children teaches them to be helpless and undermines their confidence.

◆ *Debilitating beliefs.* How mathematics is taught can profoundly affect how children view mathematics, their learning of the topic, and themselves (Baroody 1987a, 1998). When mathematics is taught in a rapid-fire, abstract, and thus incomprehensible manner, children hear such unspoken (and sometimes spoken) messages as the following: (1) "Only geniuses can understand mathematics. Just do as you're told. You're not smart enough to understand it." (2) "Mathematics is merely a bunch of facts and procedures. Normal children memorize them quickly. You're dumb if you can't." (3) "In mathematics, there is one correct method for doing things. Good children follow directions. You're bad if you use an unacceptable procedure such as counting." Such beliefs can have a powerful impact on how children go about learning and using mathematics (Cobb 1985a, 1985b; Pressley et al. 1989; Reyes 1984; Schoenfeld 1985). For example, emphasizing memorizing written procedures by rote teaches children that mathematics does not involve thinking and that their own ideas and strategies are not relevant or, at best, are inferior substitutes for real math. Because they conclude that mathematics does not make sense, many children (like Aaron) stop monitoring their work thoughtfully and dispense with assigned work as quickly as possible (e.g., Holt 1964). As a result, an unreasonable answer may not trouble them in the least, even when it is incompatible with what they do know (e.g., Hiebert and Wearne 1984; Schoenfeld 1985). For instance, children who use the small-from-large bug often overlook the fact that subtraction cannot yield a difference larger than the starting number:

$$\begin{array}{r} 22 \\ -\ 5 \\ \hline 23 \end{array}$$

Evidence of mathematical proficiency among children with learning difficulties

Is believing that mathematics instruction can foster adaptive expertise or mathematical proficiency among children with learning or behavior difficulties realistic? Is it realistic to believe that the instructional reforms that NCTM envisions (the investigative approach) would benefit such children, including those labeled learning disabled or mentally retarded?

Although we know relatively little about fostering the mathematical proficiency of children with learning difficulties (Cherkes-Julkowski and Gertner 1989; Mastropieri, Bakken, and Scruggs 1991), under the right conditions, such children are capable of at least some adaptive expertise—certainly more than previously thought possible (Baroody 1999b; Bray and Turner 1986; Ferretti 1989; Scheid 1990). For example, children with mental retardation seem capable of *understanding* basic counting principles, *discovering* and *applying* number and arithmetic rules, and *spontaneously inventing* more advanced addition strategies (Baroody 1986a, 1986b, 1987a, 1987b, 1988, 1994, 1995, 1996b; Baroody and Snyder 1983; Bråten 1996; Caycho, Gunn, and Siegal 1991; Ezawa 1996; Holcomb, Stromer, and Mackay 1997; Maydak et al. 1995).

Likewise, we know little about fostering the mathematical proficiency of children with behavioral disorders. However, recall the old teaching maxim that good instruction—meaningful, challenging, and interesting instruction—is the best method of discipline (in that it goes a long way toward preventing behavioral problems from arising). Consider Carter (Isenbargar and Baroody 2001), whose mother abandoned him when he was about four years old. This troubled child was labeled as having a behavior disorder early in his school career. Carter's school records documented fighting, swearing, disrespect toward adults, and sexual harassment. Carter aggressively challenged his fourth-grade mathematics teacher, but he responded well to her firm guidance, humor, and encouragement. The most important change came, though, when his teacher began to introduce challenging problems. Previously indifferent to mathematics, Carter found the challenge of these "cranium crackers" exciting. Now, he quickly completed the rest of his assignments to get to the fun part: solving the problems. In time, Carter recognized that his peers could be helpful in solving the cranium crackers, and he stopped trying to intimidate or upset them. In turn, his classmates recognized his problem-solving ability and became more accepting of him. In brief, Carter, who for years had struggled both academically and socially, found a measure of success and redemption in problem solving.

Why children with learning or behavioral difficulties need to achieve mathematical proficiency

Like typical developing children, special-needs children need to be taught mathematics in a way that empowers them "to use it flexibly, insightfully, and productively" (Trafton and Claus 1994, p. 19). Consider the relatively extreme case of children with moderate global developmental delays. Two trends make fostering the mathematical proficiency of such children particularly important now. One is changes in our economy. Fewer jobs involve manual labor or factory work, and more jobs are available in retailing and services. In brief, fewer jobs require the mechanical application of arithmetic (routine expertise) and more jobs require analyzing and evaluating numerical information (adaptive expertise). Also, a shift toward more self-employment or unsupervised work is evident.

A second trend is the increasing emphasis on deinstitutionalization. Individuals with moderate global developmental delays are increasingly not expected to remain institution-alized and dependent on others for their support and care. They are increasingly expected to become independent or at least semi-independent, capable of working and living in a community setting (Brown, Bellamy, and Gadberry 1971; Massey, Noll, and Stephenson 1994). As Brown, Bellamy, and Gadberry (1971) noted: "Obviously, the abilities and concepts necessary for functioning in a community setting greatly exceed those required for institutional living. . . . It is no longer acceptable for special educators to assume that [students with mental retardation] cannot learn to count, work, travel, shop, tell time, etc. The students are now in the community. We must find ways to teach them the necessary skills and concepts" (pp. 178–79).

More recently, Massey, Noll, and Stephenson (1994) noted that "parents and professionals agree that the most appropriate and desired educational outcome for students with moderate . . . retardation is the ability to participate meaningfully in the activities of daily living. Ultimately, this means living in the community and working in integrated competitive-employment settings" (p. 353). Mathematical proficiency is necessary for anyone to live and work semi-independently or independently. Only with such proficiency can individuals with moderate mental retardation achieve the necessary self-reliance to manage their personal affairs, particularly their finances (e.g., the efficient, appropriate, and flexible use of money and time skills).

How Can Teachers Maximize the Mathematical Proficiency of Children with Learning Difficulties?

How can teachers best help children with learning and behavioral difficulties learn mathematics? This section begins with general guidelines for planning and implementing effective instruction with such children. Next is a discussion of the importance of a sound developmental framework in writing effective individual educational plans (IEPs). Then I illustrate the key points of the chapter with a major bugaboo for students, particularly those with learning difficulties, namely, mastering the "basic number facts."

General guidelines for planning and implementing mathematics instruction for children with learning or behavioral difficulties

Although existing research does not furnish definitive guidelines for teaching mathematics to children with learning or behavioral difficulties, it does point to the following advice.

The instruction of children with learning or behavioral difficulties should proactively promote all four aspects of mathematical proficiency in an intertwined manner. According to NCTM's (2000) Equity Principle, "excellence in mathematics education requires equity—high expectations and strong support for all students" (p. 12). As the previous section argued, mathematical proficiency is a worthwhile and achievable goal for students with learning or behavioral difficulties. These high expectations must be supported by effective instruction that is proactive (focused on prevention of difficulties) rather than reactive (focused on remedial efforts). Using developmentally appropriate instruction can eliminate many learning difficulties and, thus, prevent a vicious circle of failure and despair.

As Kilpatrick, Swafford, and Findell in *Adding It Up* (2001) note, the four aspects of mathematical proficiency (conceptual understanding, procedural fluency, strategic and adaptive mathematical thinking, and a productive disposition) are interdependent and need to be fostered in an interwoven manner. Perhaps the best way to achieve this goal is with the investigative approach—purposeful, meaningful, and inquiry-based instruction (Baroody 1998). That is, instruction and practice account for student interests and create a real need for learning and using mathematics. It should be meaningful (i.e., focus on helping students make connections) so as to promote conceptual understanding and procedural fluency. Instruction should also be inquiry based to promote strategic and adaptive mathematical thinking. Purposeful, meaningful, and inquiry-based instruction, along with efforts to foster constructive beliefs, should promote a positive disposition among students.

The same teaching and learning principles apply to all children, including those with special needs. Research does not indicate that typically developing children learn in one way (governed by cognitive learning principles) and that children with learning or behavioral difficulties learn in a different, less sophisticated manner (governed by behavioristic "laws of learning"). Put differently: cognitive learning principles—such as NCTM's (2000) Learning Principle ("students must learn mathematics with understanding, actively building new knowledge from experience and prior knowledge" [p. 20])—apply to all children. For all children, then, instruction should build on their informal mathematical knowledge, help them to connect knowledge, prompt them to think about their knowledge, and discuss and debate their ideas and strategies. When children with learning or behavioral difficulties are taught mathematics in accordance with cognitive principles, many show significant improvement in learning concepts and skills and can exhibit at least some adaptive expertise (Baroody 1996b, 1999b).

However, cognitive principles do not indicate that children with special needs should be treated identically to their same-age peers. For example, helping children with mental retardation to construct the number or arithmetic concepts that other children do in a much shorter time may take several years (Baroody 1999b). Moreover, applying cognitive principles to teaching children with learning or behavioral difficulties may require creative adaptations or accommodations.

Instruction of children with special needs must match their knowledge and needs and should include the careful, thoughtful, and flexible use of both mainstreaming and segregated instruction. According to NCTM's (2000) Teaching Principle, "effective mathematics teaching requires understanding what students know and need to learn and then challenging and supporting them to learn it well" (p. 16). Furthermore, "there is no one 'right way' to teach" (p. 18)—to promote and support learning. Using appropriate instructional methods to meet the individual needs of children with learning or behavioral difficulties—to challenge and to support their learning—requires reflective and flexible teaching.

Consider mainstreaming. Like any instructional tool, teachers can use this method wisely or not. Currently, it is all too often used inflexibly—without regard to a student's knowledge and needs—and thus ineffectively. Consider Ann (a child with Down syndrome), who was placed in a regular eighth-grade mathematics class along with children the same age. The girl sat through class after class with little or no comprehension of the instruction. The assigned aide did try to discuss the instruction afterward but with little or no success. The aide also supplied simplified or watered-down worksheets (e.g., asking Ann what half of various amounts were instead of giving her worksheets on operations on fractions). In brief, Ann's integration into the class was in name only and did almost nothing to foster her mathematical proficiency.

Alfred Binet devised the IQ test and advocated segregated instruction for low-ability students for the most humane of reasons. As Ann's story shows, Binet saw that such children were often utterly lost in regular classrooms and suffered terribly there. Because segregated instruction was implemented poorly or abused, it was widely condemned and largely abandoned. Now educators advocate mainstreaming for the most humane of reasons. Unfortunately, this approach, too, is all too often being implemented poorly or being abused. In the end, no substitute exists for adequately supporting children with special needs. This endeavor includes offering sufficient staff who are both well trained and caring. Real improvement in the education of children with special needs will also require moving past dogmatic positions and taking a reflective approach that takes the best interest of each child into account. Doing so includes carefully and flexibly using both mainstreaming and segregated instruction. Furthermore, effective use of the former means ensuring that children with special needs are not only physically present in mainstream classrooms but also intellectually engaged in them.

Teachers need to use instructional tools such as activities, manipulatives, group work, and calculators carefully and reflectively. According to NCTM's (2000) Curriculum Principle, "a curriculum is more than a collection of activities: it must be coherent, focused on important mathematics, and well articulated across the grades" (p. 14). This statement helped frame the Council's pre-K–8 *Curriculum Focal Points for Prekindergarten through Grade 8 Mathematics: A Quest for Coherence* (NCTM 2006), which has influenced the revision of state curricular standards and arguably formed the template for the *Common Core State Standards* (Council of Chief State School Officers [CCSSO] 2010). Curriculum expectations and related classroom activities should not be carried out for their own sake but instead with clear goals in mind and a plan for achieving these goals (cf. Brownell 1935; Dewey 1963). Furthermore, planning effective (truly educational) activities requires taking into account internal factors (e.g., students' developmental readiness, needs, and interests) as well as external factors (e.g., the goals, the worthwhile task or problem presented). Teachers also need to consider how to help students reflect on and make sense of an activity.

Teachers should not assume that modeling an arithmetic procedure with manipulatives and having children imitate the model with manipulatives themselves will automatically help them understand the procedure. As Aaron's experience indicated, this is simply not the case (e.g., Baroody 1989b, 1998; Clements and McMillen 1996; Fuson and Burghardt 1993, 2003; Miura and Okamoto 2003; Seo and Ginsburg 2003). Children do not automatically see how the steps in a concrete model parallel those in a written procedure (Fuson and Burghardt 2003; Resnick 1982). They may not even understand the concrete model conceptually and may simply memorize the manipulative-based procedure by rote, as they would an incomprehensible written procedure (e.g., Baroody 1989b). In brief, instead of imposing a concrete model on students, it generally makes more sense to engage them in a meaningful problem-solving situation and to encourage them to use their existing knowledge to devise their own concrete solution method with manipulatives (see Baroody 1998 for a discussion of applying this model to operations on single-digit and multidigit whole numbers, fractions, and decimals).

According to NCTM's (2000) Technology Principle, "technology is essential in teaching and learning mathematics" (p. 24). It "offers teachers options for adapting instruction to special student needs. Students who are easily distracted may focus more intently on computer tasks, and those who have organizational difficulties may benefit from the constraints imposed by a computer environment. Students who have trouble with basic procedures can develop and demonstrate other mathematical understandings, which in turn can eventually help them learn the procedures" (p. 25). As with any instructional tool (e.g., activities, manipulatives, and grouping), a teacher cannot use technology willy-nilly and expect good learning outcomes. Teachers need to consider when and how using technology will be helpful in achieving instructional goals and meeting students' learning needs.

Instruction should be based on a child's individual pattern of informal and formal mathematical strengths and weaknesses. Effective instruction—instruction that develops mathematical proficiency—depends on effective assessment. According to NCTM's (2000) Assessment Principle, "assessment should support the learning of important mathematics and furnish useful information to both teachers and students" (p. 22).

Children should not be prejudged on the basis of a label such as slow learner, learning disabled, mentally retarded, or behaviorally disordered. Within what are presumed to be homogeneous groups of children, significant individual differences exist in the readiness and capacity to learn particular mathematical concepts, skills, and strategies (Baroody 1999b). For example, IQ did not indicate which children with mental retardation self-invented addition strategies (Baroody 1996b). Whereas some children with moderate retardation did so, some children with mild retardation and the same instruction did not.

Effective assessment identifies specific informal or formative and formal mathematical strengths and weaknesses in conceptual understanding, as well as procedural fluency (Ginsburg and Baroody 2003). This process is necessary to determine a child's developmental readiness to learn a particular concept or skill meaningfully. The next subsection further discusses the intimate relation between effective assessment or diagnostic testing and effective teaching.

Developing Individual Educational Plans

Each special education student must have an IEP. As one can see with Aaron, these plans need to focus on specific, developmentally appropriate, and sequenced competencies rather than global goals (e.g., table 2.2). This endeavor requires identifying strengths on which to build, as well as weaknesses that need to be remedied. It further requires identifying informal weaknesses and strengths, as well as formal ones—because the former are often essential for understanding the latter. IEPs need to identify how instruction will foster conceptual understanding and strategic and adaptive mathematical thinking, as well as procedural fluency. It should also include affective goals for promoting a productive disposition as well as cognitive goals.

Example: Mastering the Basic Number "Facts"

What does mastery of the basic number facts (the single-digit addition and multiplication combinations and their related subtraction and division combinations) really involve? (To avoid fostering misconceptions, I recommend using Brownell's [1935] term "number combinations" instead of the commonly misunderstood term "number facts" or "arithmetic facts." As he noted, "recognizing that $3 + 2 = 5$ is an abstraction or generalization." Whereas "arithmetic facts" connotes a particular [rote memorization] process, "combinations" does not. Furthermore, I strongly recommend that "combination fluency," or even "combination mastery," be used instead of "recall" and that it be clearly defined as including *any* efficient [fast and accurate] strategy.) How do students achieve this mastery? Why do many children with learning difficulties have tremendous problems mastering basic combinations? How can teachers best help such children to achieve computation fluency in this area?

What is basic number combination fluency, and how is it achieved?

Researchers have long debated how the basics are learned, represented in long-term memory, and processed mentally (Baroody 1985, 1994; Baroody and Tiilikainen 2003). The following summarizes two different views on these issues.

Conventional wisdom

According to conventional wisdom, mastery (efficient production) of basic number combinations entails the rapid retrieval of individually stored number facts—a process that is independent of meaningful knowledge (Ashcraft 1992; Dehaene 1997; Siegler and Jenkins 1989; Thorndike 1922). In this view, mastering basic combinations is merely a process of memorizing individual facts by rote. Students presumably can achieve this goal with enough practice and effort in relatively short order (even "dull" students, those presumably capable of little conceptual understanding). In the conventional view, relying on "immature strategies" or "crutches," such as finger counting or reasoning (e.g., 4 + 4 is 8, 5 is more than 4, so 5 + 4 is 9), *hinders* achieving the goal of fact recall. In effect, mastering the basic number combinations required filling in a mental table of facts (e.g., recreating the table of 100 basic addition facts). As conventionally conceived, then, mastery of number facts is the epitome of routine expertise.

Table 2.2
Aaron's IEPs without and with a powerful developmental framework

Without behavioral objectives	With behavioral objectives	
	Results of diagnostic testing with the TEMA-3[a] (T) and follow-up testing (FT)	Learning goals[b]
1. Correctly identify, read, and write numerals to 10 at least 95 percent of the time.	**Numeral writing**	**Numeral writing**
	Strengths. Can recognize and read numerals 0 to 9, which indicates he knows their distinguishing characteristics (parts and part–whole relations; T14 and FT). Correctly distinguishes between numerals and reverse numerals for 1–5, 7, and 8, which indicates he knows the left–right orientation of these numerals (FT).	• Needs to recognize that the loop for a 6 goes on the bottom *right* and that the loop for the 9 goes on the top *left*.
	Weaknesses. Does not distinguish between correctly oriented and reverse 6s or 9s (FT). Writes numerals 2–7 and 9 in reverse about half the time (T15 and FT).	• Needs to work on motor plan, specifically the starting point and direction for 2, 3, 7 (start at the upper left and go right), 4 (start at the upper left and go down), 5, 6, and 9 (start at the upper right and go left).
	Conclusion: Has a completely accurate mental image of the numerals 1–5, 7, and 8 and a partially accurate mental image of 6 and 9 (knows their parts and part–whole relations but not their left–right orientation). Motor plan for 1–7 and 9 does not specify the correct starting point and direction.	
2. State the sum of single-digit addition facts and the difference of corresponding subtraction facts with at least 95 percent accuracy within three seconds.	**Basic number combinations**	**Basic number combinations**
	Strengths. Can concretely solve addition and subtraction problems (e.g., can use fingers or other objects and a counting-all procedure to solve add-to or take-away word problems (T8, T16). Exhibits good mastery of verbal counting skills (T21 and T31), including automatically stating the number after another up to 20 and slowly stating the number before another up to 20 (T12, T13, and FT). Can fairly quickly decide which of two adjacent numbers is larger up to 20 (T19 and T20). Can enumerate and produce collections up to 10 with reasonable accuracy (T6, T10, T11, T23, T28).	• Needs to master the $n + 1$ and $n - 1$ number combinations by relating such combinations to his existing number-after and number-before knowledge, respectively.
		• Needs to develop more efficient informal counting strategy (counting on) for determining the sums of larger basic addition combinations to provide a more solid basis for discovering patterns and relations and, thus, inventing reasoning strategies. Once he has discovered the number-after rule for $n + 1$ combinations, encourage the self-invention of counting on by computing the sums of $n + 2$ to $n + 6$.

Table 2.2—*Continued*

Without behavioral objectives	With behavioral objectives	
	Results of diagnostic testing with the TEMA-3[a] (T) and follow-up testing (FT)	Learning goals[b]
2. State the sum of single-digit addition facts and the difference of corresponding subtraction facts with at least 95 percent accuracy within three seconds.	**Basic number combinations** *Weaknesses.* Does not know automatically the sums and differences of basic addition and subtraction combinations, not even $n + 1$ combinations (T43, T46, T50, T51). Has difficulty doing mental addition with single-digit addends (T26). Cannot count on (T32). Does not understand additive commutativity (T34). Skip counts by twos to 18 slowly and with effort. Counts backward from 10 (T24) and particularly 20 very slowly, with difficulty, and often with errors or confusion. Has difficulty solving missing-addend problems (T17). *Conclusions:* Has a conceptual understanding of addition and subtraction. Has the counting skills and concepts to support informal counting strategies for adding and subtracting. Has the counting skill, number-comparisons, number-after, and number-before knowledge to devise a reasoning strategy for $n + 1$ and $n - 1$ combinations (the number-after and number-before rule, respectively). Has the counting skill necessary to reliably add with objects or verbal counting and to subtract with objects and, thus, discover patterns and relations.	**Basic number combinations** • Encourage the discovery and discussion of reasoning strategies for addition doubles (sums form the even numbers), doubles + 1 (e.g., $5 + 4 = 4 + 4 + 1$), and so on.
3. Add and subtract up to three-digit numbers involving renaming.	**Multidigit addition and subtraction** *Strengths.* Exhibits number sense with two-digit numbers. Knows the number after a two-digit number (T36 and T39), can verbally count by tens up to 90 (T33), and understands two-digit number relations (e.g., that 32 is closer to 24 than 61; T36). Can read two-digit numerals (T35). Can (slowly) add two-digit numbers such as 23 + 15. *Weaknesses.* Has difficulty reading and writing three-digit numerals (T44 and T45). Does not understand base-ten (grouping) equivalents such as "It takes 10 tens to make a hundred." Cannot properly align "uneven" addends such as 98 − 1 and 356 − 24, or 53 + 4 and 156 + 43 (T55 and T59). Specifically, aligns digits from the left, not the right. Cannot subtract two-digit numbers involving borrowing (T69).	**Basic number combinations** • Promote an understanding of grouping by creating groups of ten and later groups of 100. Use interlocking blocks that can be composed into ten sticks (which can then be decomposed into tens units). Learn Egyptian hieroglyphics for representing multidigit numbers (e.g., $234 \rightarrow$ ◎◎∩∩ ∩|||| ; $204 \rightarrow$ ◎◎||||). • Next, promote an understanding of place value by representing hundreds, tens, and ones and groups of ten and ones by using a place-value mat ($\overline{100|10|1}$).

Table 2.2—*Continued*

Without behavioral objectives	With behavioral objectives	
	Results of diagnostic testing with the TEMA-3[a] (T) and follow-up testing (FT)	Learning goals[b]
3. Add and subtract up to three-digit numbers involving renaming.	*Multidigit addition and subtraction*	*Multidigit addition and subtraction*
	Conclusions. Has a counting- or unit-based understanding of two-digit numbers, not grouping-by-ten, place-value conceptions (e.g., views 23 as 23 units, not also as 2 tens and 3 ones). This lack of conceptual understanding contributes to his difficulty with reading and writing three-digit numerals. Because he does not understand the grouping-by-ten concept, base-ten equivalents such as 10 ones = 1 ten have no meaning for him and have not been learned. Likewise, his alignment difficulties stem from the fact that he does not recognize that only digits of the same place value can be added or subtracted—for example, the 5 and 4 in the expression 156 + 43 both represent groups of ten and, thus, the 4 (tens) must be combined with the 5 (tens), not the 1 (one hundred). Likewise, without understanding grouping-by-ten and place-value concepts, one cannot understand carrying and borrowing.	For example, 135 would be represented by one block on the hundreds section, three blocks (of the same color) on the tens section, and five blocks on the ones section. Encourage to label each section with a written digit number to represent the multidigit number (e.g., 1 3 5). • Encourage learning of base-ten equivalents (e.g., 10 ones = 1 ten; 10 tens = 1 hundred) and other equivalents (e.g., 42 can be represented by 42 ones or 42 cubes, 4 tens and 2 ones or 4 longs and 2 cubes, or 3 tens and 12 ones or 3 longs and 12 cubes). • Encourage to devise a procedure that involves base-ten blocks or Egyptian hieroglyphics for solving two-digit addition without and with carrying and subtraction without and with borrowing. • Encourage to devise a written procedure that models each step of the block model or the hieroglyphics model.

[a]See Ginsburg and Baroody (2003).

[b]See, for example, Baroody (1998) for general guidelines; content-specific instructional suggestions; and examples of activities, projects, games, children's literatures, technology, or manipulatives for achieving these learning goals. See chapter 5 for illustrative vignettes.

An alternative: The number sense view

Mastery of (fluency with) basic number combinations or basic facts is more complicated than conventional wisdom suggests. First of all, even adults use a variety of methods, including fast counting or efficient reasoning strategies, to accurately and quickly determine answers to basic combinations (Browne 1906; LeFevre, Bisanz, et al. 1996; LeFevre, Sadesky, and Bisanz 1996; see LeFevre et al. 2003 for a review). Second, number sense in the form of relational knowledge (meaningful connections) may affect how basic number-combination knowledge is mentally represented in long-term memory and processed (Baroody 2006; Gersten and Chard 1999; Jordan 2007). For example, an understanding of commutativity may allow us to store $5 \times 8 = 40$ and $8 \times 5 = 40$ as a single triplet: 5, 8, 40 (Baroody 1999c; Butterworth, Marschesini, and Girelli 2003; Rickard and Bourne 1996; Rickard, Healey, and Bourne 1994; Sokol et al. 1991). In brief, the basic number combinations may not be represented as a table of facts but as a network of facts and interconnecting relations (e.g., Baroody 1985, 1987a, 1994). The observation that the computational prowess of idiots savants does not stem from a rich store of isolated facts but from a rich number sense (Heavey 2003) further supports this view.

Recent research (e.g., Baroody 1999a, 1999c) confirms earlier work that mastering the basic number combinations is a slow process (e.g., Brownell 1935), one that proceeds through three phases: (1) relatively slow counting strategies, (2) relatively slow reasoning strategies, and (3) mastery (relatively fast fact retrieval or nonretrieval processes [e.g., some automatic or automatic reasoning strategies or rules; Baroody 1998]). In the long run, though, meaningful learning, including the discovery of patterns or relations, can greatly facilitate mastering basic number combinations (Baroody 1985, 1994, 1999a, 1999c; Brownell 1935; Rathmell 1978). Allowing children to use counting or reasoning strategies is important for mastering the basic combinations for several reasons:

◆ Using informal strategies can lead to discovering patterns and relations that underlie an efficient mental representation and processing of basic combinations. Consider two examples. (1) Recognizing the pattern that "adding 0 to any number doesn't change it" can serve as a rule for generating the answer to an infinite number of $n + 0$ and $0 + n$ combinations (Baroody 1989a, 1992). (2) Discovering the relation between existing counting knowledge, specifically their number-after knowledge, and $n + 1$ and $1 + n$ combinations allows children to answer the former efficiently (quickly and accurately; Baroody 1989a, 1992, 1995). According to the resulting number-after rule for $n + 1$ combinations, $7 + 1$, for instance, is simply the number after *seven* in the counting sequence, namely, *eight*.

◆ Practice using (relatively slow) counting and reasoning strategies will, in time, lead to their efficient execution. For instance, with practice, applying the number-after rule for $n + 1$ combinations becomes routine.

◆ In the long run, a meaningful approach is a more efficient way to foster combination mastery. Although this meaningful learning of basic number combinations may initially be relatively slow, it ultimately requires less time and effort than does the traditional approach. It is easier to learn and remember meaningful and well-connected facts than meaningless and isolated ones for two reasons. (1) Meaningful memorization facilitates transfer to nonpracticed combinations and, thus, reduces the practice needed to master combinations. For example, once children discover the number-after rule for $n + 1$ combinations, they can readily apply this rule to any $n + 1$ combination for which they know the counting rules to generate a next number—including previously unpracticed single-digit or even multidigit $n + 1$ combinations (Baroody 1989a, 1992, 1995). Similarly, knowledge of the multiplicative commutativity principle reduces by about half the number of multiplication combinations that a child must memorize. For instance, if children know that 8×6 is 48 and that the order of the factors does not affect the product, they can quickly reason (deduce) that the product of the unpracticed and unknown combination 6×8 is 48 also (Baroody 1999c). (2) Meaningful learning of combinations also results in better retention (less forgetting) and, thus, reduces the amount of review necessary later.

◆ Unlike the conventional approach, a meaningful approach to teaching the basic facts or basic number combinations is far more likely to achieve all four aspects of mathematical proficiency and to do so in an intertwined manner. (Table 2.3 compares the two views of number-combination mastery in the context of these aspects.) Allowing children to use informal counting and then reasoning strategies enables them to determine answers in a manner that makes sense to them. By using strategies that they trust and understand, children are more likely to apply inductive reasoning to discover patterns and relations, use deductive reasoning to use what they know to figure out what they do not know, and achieve adaptive expertise with basic combinations—procedural fluency characterized by efficient (accurate and fast), appropriate, and flexible use of this basic knowledge. For example, by discovering that the doubles—such as $4 + 4 = 8$, $5 + 5 = 10$, $6 + 6 = 12$, and $7 + 7 = 14$—result in sums that parallel the even numbers (skip counting by two), children can more quickly and securely commit these combinations to memory (Baroody 1998). This example shows how making a connection (conceptual insight or discovering a relation) can promote procedural fluency. In turn, this conceptually based fluency with the addition doubles allows children to engage in strategic and adaptive mathematical thinking (e.g., discovering new patterns or relations), which leads to further conceptual understanding, strategic and adaptive thinking, and procedural fluency. For instance, they may recognize that "near doubles" such as $6 + 7$ must have a sum that falls between the even numbers (i.e., have an odd sum) and use this insight to logically reason out unknown sums (i.e., the sum of $6 + 7$ is one more than that of $6 + 6$ or one less than the sum of $7 + 7$). As this reasoning becomes more automatic, computational fluency increases. The by-product of this interesting, meaningful, and inquiry-based process for achieving procedural fluency is likely to be a productive disposition. For example, it can help children better understand the true nature of mathematics (a search for patterns in order to solve problems), view mathematics as sensible, and believe in their own efficacy. For instance, a meaningful approach is not only more interesting and more effective but also more pleasant.

According to the alternative view, then, mastery of the basic number combinations should epitomize adaptive expertise. Mastering these combinations is not merely a process of filling in a mental number-fact table but a process of building understanding, of seeing relations or connections. In a real sense, it should be viewed as a by-product of constructing a rich number sense.

Why do children with learning difficulties have problems with combination mastery?

Many children with mathematical learning difficulties "have problems remembering basic arithmetic facts, such as $5 + 9 = 14$, even with extensive drilling" (Geary 1996). A persistent "fact-retrieval deficit" is a common characterization of such children (for reviews, see Allardice and Ginsburg 1983; Geary 1996; Jordan et al. 2003; Mastropieri et al. 1991; Scheid 1990). Less clear is why children with learning difficulties have problems in this area. Next I outline two views on the source of difficulties with basic number combinations.

Cognitive-deficit view

Geary (1996) concluded that "many children have difficulties learning basic arithmetic because of a serious cognitive or neuropsychological disorder, certain forms of which might be heritable." A key symptom—a persistent lack of number-combination fluency—he characterized as a "developmental difference" rather than a "developmental delay." This difference, Geary proposed, may involve a memory deficit (poor working-memory skills), slow mental processing, or a neurological dysfunction in brain regions that involve storing the number facts. In effect, "fact-retrieval deficits" are attributed to an inability to achieve routine expertise with basic combinations.

Developmentally inappropriate instruction view

Although some children who have difficulty mastering basic number combinations may well have cognitive deficits, probably most are simply the victims of developmentally inappropriate instruction. That is, their instruction does not encourage—and even discourages—the informal bases for combination fluency, namely, the use of self-invented and meaningful counting and reasoning strategies. Put differently: the arithmetic instruction of children with learning difficulties typically does not focus on sense making and looking for patterns and relations. As a result, they do not develop the adaptive expertise (conceptual understanding and flexibility) that characterized experts' combination fluency. Moreover, such instruction may well contribute to the learning difficulties of children with real cognitive deficits.

What evidence supports the instructional deficit view? Most previous research with children with learning difficulties has focused on memorizing arithmetic facts by rote—that is, skipping the second (reasoning) phase and sometimes even the first (counting)

Table 2.3
Two views of number-combination mastery and its impact on mathematical proficiency

Aspect of mathematical proficiency	Conventional view (direct-instruction-and-drill approach)	NCTM Standards–based view (investigative approach)
Conceptual understanding	Moving directly to using flashcard drills, "around the world," worksheets, and other methods of practice to memorize facts by rote does little, if anything, to promote conceptual understanding.	Conceptual or meaningful groundwork for an operation must be laid first. During this time, children are encouraged to invent and use counting strategies. During the reasoning phase, children are encouraged to look for patterns and relations to devise reasoning strategies. Doing so can entail relating unfamiliar or unknown combinations to familiar or known combinations (assimilation or integration of knowledge).
Procedural fluency	Memorizing facts by rote may allow children to reproduce them efficiently (quickly and accurately) and even appropriately. However, it may also result in forgetting or confusion (e.g., responding to $n \times 0$ combinations such as 7×0 with the sum of the numbers)—either of which reduces efficiency. Children may also fail to recognize many appropriate applications of such knowledge. Routine expertise with combinations is not likely to be adapted or used flexibly.	Meaningful memorization (memorization based on understanding and built on forging connections) makes mastering, retaining, and applying this knowledge easier. That is, adaptive expertise with basic combinations makes it more likely to achieve efficiency, appropriate use, and flexible use.
Strategic and adaptive mathematical thinking	Memorizing facts by rote does little, if anything, to develop problem-solving, reasoning, representing, or communicating abilities. Indeed, this method will probably interfere with children's development of inquiry competencies.	Meaningful memorization of basic number combinations involves children in inductive reasoning: looking for patterns and relations and formulating reasoning strategies. It also involves them in deductive reasoning: using logical reasoning and what they know to figure out unknown combinations. Sharing reasoning strategies and their justification involves representing and communicating. Reasoning out or recalling basic number combinations in the context of solving problems offers practice in problem solving, as well as combination practice.
Productive disposition	Memorizing basic number facts by rote does not help children appreciate the value and true nature of mathematics. Timed tests and an overemphasis on responding quickly may help create beliefs that mathematics is something that they must do in a hurry. Moreover, it may create fear of mathematics or even math anxiety.	Meaningful memorization of basic number combinations reinforces the essential point that the heart of mathematics is looking for patterns and relations to solve problems. Self-discovery of patterns or relation and the self-invention of reasoning strategies can create confidence in one's own mathematical ability.

phase (Broome and Wambold 1977; Grim, Bijou, and Parsons 1973; Kokaska 1985; Miller 1976; and Ogletree and Ujlaki 1976—all cited in Mastropieri, Bakken, and Scruggs 1991). Although children with learning difficulties have been taught reasoning strategies through direct instruction (e.g., Thornton and Smith 1988), some evidence suggests that such children cannot invent these strategies spontaneously (Swanson and Cooney 1985; Swanson and Rhine 1985). There is reason to believe, though, that such children can benefit from instruction that focuses on looking for patterns and relations and using these regularities to devise reasoning strategies (e.g., Myers and Thornton 1977; Thornton 1978; Thornton, Jones, and Toohey 1983; Thornton and Toohey 1985). Even children with severe learning difficulties are capable of at least some aspects of adaptive expertise with basic combinations (Baroody 1999b). For example, with meaningful instruction, children with mental retardation can *discover* the $n + 0 = n$ rule and the number-after rule for $n + 1$ combinations and then *apply* these rules efficiently and effectively with previously unpracticed and unknown combinations (e.g., Baroody 1988, 1995).

How can teachers best help children with learning difficulties to master basic number facts and combinations?

◆ Because "experts" use a variety of strategies, including automatic or semiautomatic rules and reasoning processes, we should define number combination proficiency or mastery broadly as including *any efficient strategy*, not narrowly as fact retrieval. Thus, students should be encouraged, not discouraged, from flexibly using a variety of strategies.

◆ Teachers should patiently help children construct number sense and not prematurely drill facts. That is, teachers should encourage the invention, sharing, and refinement of informal strategies. Children typically adopt more efficient strategies as their number sense expands or when a real need to do so surfaces (e.g., in determining an outcome of rolling dice in an interesting game).

◆ To promote meaningful memorization or mastery of basic combinations (combination fluency), teachers should *focus on encouraging children to look for patterns and relations*; to use these discoveries to construct reasoning strategies; and to share, justify, and discuss their strategies. A direct implication of this guideline is that children should concentrate on "fact families," not individual facts (see box 5.6 on pp. 5-31–5-33 of Baroody [1998] for a discussion of the developmental bases and learning of these fact families).

◆ To maximize combination fluency and other aspects of mathematical proficiency, teachers should encourage children *to build on what they already know*. For example, the sum of $n + 1$ combinations is simply the number after n in the counting sequence; $5 - 3$ can be thought of as $3 + ? = 5$; and $2 \times n$ combinations are equal to the addition doubles (for empirical support for each example, see Baroody 1995, 1999a, and 1993, respectively).

◆ Different reasoning strategies may require different approaches. Patterns and relations differ in their salience (Baroody 1999a; Baroody, Ginsburg, and Waxman 1983). Unguided discovery learning might be appropriate for highly salient patterns or relations, such as additive commutativity. Less obvious patterns may require more structured discovery learning activities, such as the complementary relations between addition and subtraction (e.g., 9 – 5 can be thought of as 5 + ? = 9; Baroody 1999a).

◆ Practice should focus on making reasoning strategies more automatic, not on drilling isolated facts.

◆ The learning and practice of number combinations should be done purposefully (see Baroody 1998 for examples). Purposeful practice is more effective than drill and practice. For example, when instruction focused on problem solving, children not only became better problem solvers but also mastered more combinations than did children whose instruction focused on drill and practice of basic facts (Carpenter et al. 1989).

Conclusions

Mathematical proficiency (conceptual understanding, procedural fluency, strategic and adaptive mathematical thinking, and a productive disposition) and the adaptive expertise (the ability to appropriately and flexibly apply meaningful knowledge to new tasks or problems) that it embodies are becoming more and more important in our technology- and information-based society. Achieving such proficiency or expertise in primary-level mathematics is essential for the meaningful learning and successful use of later school mathematics, such as fractions, decimals, and percent, and applying school mathematics to everyday life, including money management (e.g., determining discounts and balancing a checkbook). In effect, developing this proficiency early provides the cognitive foundations necessary to develop the mathematical proficiency or adaptive expertise needed for functioning independently in our complex society. Fostering the adaptive expertise of students with learning or behavioral difficulties at all levels of school is particularly important now, with the increasing emphasis on deinstitutionalization. Only with adaptive expertise will such individuals have the necessary self-reliance to manage their personal affairs, particularly their finances.

Helping children with learning or behavioral difficulties achieve mathematical proficiency and adaptive expertise will require an approach different from and more sophisticated than the traditional direct-instruction-and-drill method. It will require purposeful, meaningful, and inquiry-based instruction that promotes all aspects of mathematical proficiency in an integrated manner. Planning and implementing such instruction will require considerable time, effort, and knowledge from teachers, but significantly greater student development will reward their investment.

(Preparation of the manuscript was supported, in part, by grants from the National Science Foundation [BCS-0111829, "Foundations of Number and Operation Sense"] and the U.S. Department of Education [R305K050082, "Developing an Intervention to Foster Early Number Sense and Skill"]. The opinions expressed are solely those of the author and do not necessarily reflect the position, policy, or endorsement of the National Science Foundation or the U.S. Department of Education.)

References

Allardice, Barbara S., and Herbert P. Ginsburg. "Children's Learning Problems in Mathematics." In *The Development of Mathematical Thinking*, edited by Herbert P. Ginsburg, pp. 319–49. New York: Academic Press, 1983.

Ashcraft, Mark H. "Cognitive Arithmetic: A Review of Data and Theory." *Cognition* 44 (August 1992): 75–106.

Atwater, Jane B., Judith J. Carta, Ilene S. Schwartz, and Scott R. McConnell. "Blending Developmentally Appropriate Practice and Early Childhood Special Education: Redefining Best Practice to Meet the Needs of All Children." In *Diversity and Developmentally Appropriate Practices*, edited by Bruce L. Mallory and Rebecca S. New, pp. 1–14. New York: Teachers College Press, 1994.

Baroody, Arthur J. "Mastery of the Basic Number Combinations: Internalization of Relationships or Facts?" *Journal for Research in Mathematics Education* 16 (March 1985): 83–98.

———. "Basic Counting Principles Used by Mentally Retarded Children." *Journal for Research in Mathematics Education* 17 (November 1986a): 382–89.

———. "Counting Ability of Moderately and Mildly Mentally Handicapped Children." *Education and Training of the Mentally Retarded* 21 (December 1986b): 289–300.

———. *Children's Mathematical Thinking: A Developmental Framework for Preschool, Primary, and Special Education Teachers.* New York: Teachers College Press, 1987a.

———. "Problem Size and Mentally Retarded Children's Judgment of Commutativity." *American Journal of Mental Deficiency* 91 (January 1987b): 439–42.

———. "Mental-Addition Development of Children Classified as Mentally Handicapped." *Educational Studies in Mathematics* 19 (August 1988): 369–88.

———. "Kindergartners' Mental Addition with Single-Digit Combinations." *Journal for Research in Mathematics Education* 20 (March 1989a): 159–72.

———. "One Point of View: Manipulatives Don't Come with Guarantees." *Arithmetic Teacher* 37 (October 1989b): 4–5.

———. "The Development of Kindergartners' Mental-Addition Strategies." *Learning and Individual Differences* 4, no. 3 (1992): 215–35.

———. "Early Mental Multiplication Performance and the Role of Relational Knowledge in Mastering Combinations Involving Two." *Learning and Instruction* 3, no. 2 (1993): 93–111.

———. "Self-Regulated Learning and Use of Addition Strategies by Children Classified as Mentally Handicapped." Paper presented at the annual meeting of the American Educational Research Association, New Orleans, April 1994.

———. "The Role of the Number-After Rule in the Invention of Computational Shortcuts." *Cognition and Instruction* 13, no. 2 (1995): 189–219.

———. "An Investigative Approach to Teaching Children Labeled 'Learning Disabled.'" In *Cognitive Approaches to Learning Disabilities*. 3rd ed. Edited by D. Kim Reid, Wayne P. Hresko, and H. Lee Swanson, pp. 545–615. Austin, Tex.: Pro-Ed, 1996a.

———. "Self-Invented Addition Strategies by Children Classified as Mentally Handicapped." *American Journal of Mental Retardation* 101 (July 1996b): 72–89.

———. *Fostering Children's Mathematical Power: An Investigative Approach to K–8 Mathematics Instruction.* With Ronald T. Coslick. Mahwah, N.J.: Lawrence Erlbaum Associates, 1998.

———. Children's Relational Knowledge of Addition and Subtraction. *Cognition and Instruction* 17, no. 2 (1999a): 137–75.

———. "The Development of Basic Counting, Number, and Arithmetic Knowledge among Children Classified as Mentally Retarded." In *International Review of Research in Mental Retardation, Vol. 22,* edited by Laraine M. Glidden, pp. 51–103. New York: Academic Press, 1999b.

———. "The Roles of Estimation and the Commutativity Principle in the Development of Third Graders' Mental Multiplication." *Journal of Experimental Child Psychology* 74 (November 1999c): 157–93.

———. "Why Children Have Difficulties Mastering the Basic Number Facts and How to Help Them." *Teaching Children Mathematics* 13 (August 2006): 22–31.

Baroody, Arthur J., Neet Priya Bajwa, and Michael Eiland. "Why Can't Johnny Remember the Basic Facts?" *Developmental Disabilities Research Reviews* 15, no. 1 (2009): 69–79.

Baroody, Arthur J., and Ann Dowker. *The Development of Arithmetic Concepts and Skills: Constructing Adaptive Expertise.* Mahwah, N.J.: Lawrence Erlbaum Associates, 2003.

Baroody, Arthur J., Yingying Feil, and Amanda R. Johnson. "An Alternative Reconceptualization of Procedural and Conceptual Knowledge." *Journal for Research in Mathematics Education* 38 (March 2007): 115–31.

Baroody, Arthur J., and Herbert P. Ginsburg. "A Cognitive Approach to Assessing the Mathematical Difficulties of Children Labeled Learning Disabled." In *Handbook on the Assessment of Learning Disabilities: Theory, Research, and Practice,* edited by H. Lee Swanson, pp. 177–227. Austin, Tex.: Pro-Ed, 1991.

Baroody, Arthur J., Herbert P. Ginsburg, and Barbara Waxman. "Children's Use of Mathematical Structure." *Journal for Research in Mathematics Education* 14 (May 1983): 156–68.

Baroody, Arthur J., and Patricia M. Snyder. "A Cognitive Analysis of Basic Arithmetic Abilities of TMR Children." *Training of the Mentally Retarded* 18 (December 1983): 253–59.

Baroody, Arthur J., and Sirpa Tiilikainen. "Two Perspectives on Addition Development." In *The Development of Arithmetic Concepts and Skills: Constructing Adaptive Expertise,* edited by Arthur J. Baroody and Ann Dowker, pp. 75–125. Mahwah, N.J.: Lawrence Erlbaum Associates, 2003.

Berch, Daniel B., and Michèle M. M. Mazzocco, eds. *Why Is Math So Hard for Some Children? The Nature and Origins of Mathematical Learning Difficulties and Disabilities.* Baltimore: Paul H. Brookes Publishing Co., 2007.

Bråten, Ivar. *Cognitive Strategies in Mathematics.* Report No. 10. Oslo, Norway: Institute for Educational Research, University of Oslo, 1996.

Bray, Norman W., and Lisa A. Turner. "The Rehearsal Deficit Hypothesis." In *International Review of Research in Mental Retardation, Vol. 14,* edited by Norman R. Ellis and Norman W. Bray, pp. 47–71. Orlando, Fla.: Academic Press, 1986.

Bredekamp, Sue. "The Relationship between Early Childhood Education and Early Childhood Special Education: Healthy Marriage or Family Feud?" *Topics in Early Childhood Special Education* 13 (Fall 1993): 258–73.

Brown, Ann L., John D. Bransford, Robert A. Ferrara, and Joseph C. Campione. "Learning, Remembering, and Understanding." In *Handbook of Child Psychology.* 4th ed. Edited by John H. Flavell and Ellen M. Markman, pp. 77–166. New York: Wiley, 1983.

Brown, Lou, T. Bellamy, and E. Gadberry. "A Procedure for the Development and Measurement of Rudimentary Quantitative Concepts in Low Functioning Trainable Students." *Training School Bulletin* 68 (1971): 178–85.

Browne, Charles E. "The Psychology of Simple Arithmetical Processes: A Study of Certain Habits of Attention and Association." *American Journal of Psychology* 17 (January 1906): 2–37.

Brownell, William A. "Psychological Considerations in the Learning and the Teaching of Arithmetic." In *The Teaching of Arithmetic*, Tenth Yearbook of the National Council of Teachers of Mathematics (NCTM), edited by W. D. Reeve, pp. 1–50. New York: Bureau of Publications, Teachers College, Columbia University, 1935.

Butterworth, Brian, Noemi Marchesini, and Luisa Girelli. "Basic Multiplication Combinations: Passing Storage or Dynamic Reorganization." In *The Development of Arithmetic Concepts and Skills: Constructing Adaptive Expertise*, edited by Arthur J. Baroody and Ann Dowker, pp. 189–202. Mahwah, N.J.: Lawrence Erlbaum Associates, 2003.

Carnegie Forum on Education and the Economy. *A Nation Prepared: Teachers for the 21st Century.* New York: Carnegie Corp., 1986.

Carpenter, Thomas P., Elizabeth Fennema, Penelope L. Peterson, Chi-Pang Chiang, and Megan Loef. "Using Knowledge of Children's Mathematics Thinking in Classroom Teaching: An Experimental Study." *American Educational Research Journal* 26, no. 4 (1989): 499–531.

Carta, Judith J. "Developmentally Appropriate Practice: A Critical Analysis as Applied to Young Children with Disabilities." *Focus on Exceptional Children* 27 (April 1995): 1–14.

Carta, Judith J., Ilene S. Schwartz, Jane B. Atwater, and Scott R. McConnell. "Developmentally Appropriate Practice: Appraising Its Usefulness for Young Children with Disabilities." *Topics in Early Childhood Special Education* 11 (Spring 1991): 1–20.

Caycho, L., P. Gunn, and M. Siegal. "Counting by Children with Down Syndrome." *American Journal on Mental Retardation* 95 (March 1991): 575–83.

Cherkes-Julkowski, Miriam, and Nancy Gertner. *Spontaneous Cognitive Processes in Handicapped Children.* New York: Springer, 1989.

Clements, Douglas H., and Michael T. Battista. "Geometry and Spatial Reasoning." In *Handbook of Research on Mathematics Teaching and Learning*, edited by Douglas A. Grouws, pp. 420–64. New York: Macmillan, 1992.

Clements, Douglas H., and Sue McMillen. "Rethinking 'Concrete' Manipulatives." *Teaching Children Mathematics* 2 (January 1996): 270–79.

Clements, Douglas, and Julie Sarama. *Engaging Young Children in Mathematics: Findings of the National 2000 Conference for Preschool and Kindergarten Mathematics Education.* Mahwah, N.J.: Lawrence Erlbaum Associates, 2004.

Cobb, Paul. "A Reaction to Three Early Number Papers." *Journal for Research in Mathematics Education* 16 (March 1985a): 141–45.

———. "Two Children's Anticipations, Beliefs, and Motivations." *Educational Studies in Mathematics* 16 (May 1985b): 111–26.

Coles, Gerald S. "The Learning Disabilities Test Battery: Empirical and Social Issues." *Harvard Educational Review* 48 (August 1978): 313–40.

Council of Chief State School Officers (CCSSO). *Common Core State Standards.* Washington, D.C.: CCSSO, 2010. www.corestandards.org.

Court, Sophie R. A. "Numbers, Time, and Space in the First Five Years of a Child's Life." *Pedagogical Seminary* 27 (March 1920): 71–89.

Davis, Robert B. *Learning Mathematics: The Cognitive Science Approach to Mathematics Education.* Norwood, N.J.: Ablex, 1984.

Dehaene, Stanislas. *The Number Sense: How the Mind Creates Mathematics.* New York: Oxford University Press, 1997.

Delazer, Margarete. "Neuropsychological Findings on Conceptual Knowledge of Arithmetic." In *The Development of Arithmetic Concepts and Skills: Constructing Adaptive Expertise*, edited by Arthur J. Baroody and Ann Dowker, pp. 385–407. Mahwah, N.J.: Lawrence Erlbaum Associates, 2003.

DeRuiter, James A., and William L. Wansart. *Psychology of Learning Disabilities*. Rockville, Md.: Aspen, 1982.

Dewey, John. *Experience and Education*. New York: Collier, 1963.

Ezawa, B. *Zdhlen and rechnen bei geistig behinderten schülern: Leistungen, Konzepte und strategien junger erwachsener mit hirnfunktionsstdrungen* [*Counting and Calculating of Students with Mental Retardation: Capabilities, Concepts, and Strategies of Young Adults with Brain Dysfunction*]. Frankfurt, Germany: Peter Lang, 1996.

Ferretti, Ralph P. "Problem Solving and Strategy Production in Mentally Retarded Persons." *Research in Developmental Disabilities* 10, no. 1 (1989): 19–31.

Ferretti, Ralph P., and Albert R. Cavalier. "Constraints on the Problem Solving of Persons with Mental Retardation." In *International Review of Research in Mental Retardation, Vol. 17*, edited by Norman W. Bray, pp. 153–92. San Diego: Academic Press, 1991.

Fuson, Karen C., and Birch H. Burghardt. "Group Case Studies of Second Graders Inventing Multidigit Addition Procedures for Base-Ten Blocks and Written Marks." In *Proceedings of the Fifteenth Annual Meeting of the North American Chapter of the International Group for the Psychology of Mathematics Education*, edited by Joanne R. Becker and Barbara J. Pence, pp. 1-240–1-246. San Jose, Calif.: Center for Mathematics and Computer Science Education, 1993.

———. "Multidigit Addition and Subtraction Methods Invented in Small Groups and Teacher Support of Problem Solving and Reflection." In *The Development of Arithmetic Concepts and Skills: Constructing Adaptive Expertise*, edited by Arthur J. Baroody and Ann Dowker, pp. 267–304. Mahwah, N.J.: Lawrence Erlbaum Associates, 2003.

Geary, David C. *Children's Mathematical Development: Research and Practical Applications*. Washington, D.C.: American Psychological Association, 1996.

Gersten, Russell, and David Chard. "Number Sense: Rethinking Arithmetic Instruction for Students with Mathematical Disabilities." *Journal of Special Education* 33 (Spring 1999): 18–28.

Ginsburg, Herbert P. *Children's Arithmetic*. New York: D. Van Nostrand, 1977.

Ginsburg, Herbert P., and Arthur J. Baroody. *Test of Early Mathematics Ability*. 3rd ed. Austin, Tex.: Pro-Ed, 2003.

Ginsburg, Herbert P., Alice Klein, and Prentice Starkey. "The Development of Children's Mathematical Knowledge: Connecting Research with Practice." In *Handbook of Child Psychology: Vol. 4. Child Psychology in Practice*. 5th ed. Edited by Irving E. Sigel and K. Ann Renninger, pp. 401–76. New York: Wiley and Sons, 1998.

Giordano, Gerard. "Fourth Invited Response: The NCTM Standards: A Consideration of the Benefits." *Remedial and Special Education,* 14 (November–December 1993): 28–32.

Hatano, Giyoo. "Social and Motivational Bases for Mathematical Understanding." In *Children's Mathematics*, edited by Geoffrey B. Saxe and Maryl Gearhart, pp. 55–70. San Francisco: Jossey-Bass, 1988.

Heavey, Lisa. "Arithmetical Savants." In *The Development Of Arithmetic Concepts and Skills: Constructing Adaptive Expertise*, edited by Arthur J. Baroody and Ann Dowker, pp. 409–33. Mahwah, N.J.: Lawrence Erlbaum Associates, 2003.

Hendrickson, A. Dean. "Prevention or Cure? Another Look at Mathematics Learning Problems." In *Interdisciplinary Voices in Learning Disabilities and Remedial Education*, edited by Douglas Carnine, David Elkind, A. Dean Hendrickson, Donald Meichenbaum, Robert L. Sieben, and Frank Smith, pp. 93–107. Austin, Tex.: Pro-Ed, 1983.

Hiebert, James, and Thomas P. Carpenter. "Learning and Teaching with Understanding." In *Handbook of Research on Mathematics Teaching and Learning*, edited by Douglas Grouws, pp. 65–97. New York: Macmillan, 1992.

Hiebert, James, and Diana Wearne. "A Model of Students' Decimal Computation Procedures." Paper presented at the annual meeting of the National Council of Teachers of Mathematics, San Francisco, April 1984.

Hofmeister, Alan M. "Elitism and Reform in School Mathematics." *Remedial and Special Education* 14 (November–December 1993): 8–13.

Holcomb, William L., Robert Stromer, and Harry A. Mackay. "Transitivity and Emergent Sequence Performances in Young Children." *Journal of Experimental Child Psychology* 65 (April 1997): 96–124.

Holt, John C. *How Children Fail*. New York: Delta, 1964.

Isenbarger, Lynn M., and Arthur J. Baroody. "Fostering the Mathematical Power of Children with Behavioral Difficulties: The Case of Carter." *Teaching Children Mathematics* 7 (April 2001): 468–71.

Jordan, Nancy C. "The Need for Number Sense." *Educational Leadership* 65 (October 2007): 63–66.

Jordan, Nancy, Laurie B. Hanich, and Heather Z. Uberti. "Mathematical Thinking and Learning Difficulties." In *The Development of Arithmetic Concepts and Skills: Constructing Adaptive Expertise*, edited by Arthur J. Baroody and Ann Dowker, pp. 359–83. Mahwah, N.J.: Lawrence Erlbaum Associates, 2003.

Kaliski, Lotte. "Arithmetic and the Brain-Injured Child." *Arithmetic Teacher* 9 (May 1962): 245–51.

Kilpatrick, Jeremy. "A Retrospective Account of the Past Twenty-Five Years of Research on Teaching Mathematical Problem Solving." In *Teaching and Learning Mathematical Problem Solving*, edited by Edward A. Silver, pp. 1–15. Hillsdale, N.J.: Lawrence Erlbaum Associates, 1985.

Kilpatrick, Jeremy, Jane Swafford, and Bradford Findell, eds. *Adding It Up: Helping Children Learn Mathematics*. Washington, D.C.: National Academies Press, 2001.

LeFevre, Jo-Anne, Jeffrey Bisanz, Karen E. Daley, Lisa Buffone, Stephanie L. Greenham, and Greg S. Sadesky. "Multiple Routes to Solution of Single-Digit Multiplication Problems." *Journal of Experimental Psychology: General* 125 (September 1996): 284–306.

LeFevre, Jo-Anne, Greg S. Sadesky, and Jeffrey Bisanz. "Selection of Procedures in Mental Addition: Reassessing the Problem Size Effect in Adults." *Journal of Experimental Psychology: Learning, Memory, and Cognition* 22 (January 1996): 216–30.

LeFevre, Jo-Anne A., Brenda L. Smith-Chant, Karen Hiscock, Karen E. Daley, and Jason Morris. "Young Adults' Strategic Choices in Simple Arithmetic: Implications for the Development of Mathematical Representations." In *The Development of Arithmetic Concepts and Skills: Constructing Adaptive Expertise*, edited by Arthur J. Baroody and Ann Dowker, pp. 203–28. Mahwah, N.J.: Lawrence Erlbaum Associates, 2003.

Mallory, Bruce L. "Inclusive Policy, Practice, and Theory for Young Children with Developmental Differences." In *Diversity and Developmentally Appropriate Practices*, edited by Bruce L. Mallory and Rebecca S. New, pp. 44–62. New York: Teachers College Press, 1994.

Massey, Ann, Mary Beth Noll, and Joanne Stephenson. "Spatial Sense and Competitive-Employment Options for Students with Mental Retardation." In *Windows of Opportunity: Mathematics for Students with Special Needs*, edited by Carol A. Thornton and Nancy S. Bley, pp. 353–65. Reston, Va.: National Council of Teachers of Mathematics, 1994.

Mastropieri, Margo A., Jeffrey P. Bakken, and Thomas E. Scruggs. "Mathematics Instruction for Individuals with Mental Retardation: A Perspective and Research Synthesis." *Education and Training of the Mentally Retarded* (June 1991): 115–29.

Maydak, Michael, Robert Stromer, Harry A. Mackay, and Lawrence T. Stoddard. "Stimulus Classes in Matching to Sample and Sequence Production: The Emergence of Numeric Relations." *Research in Developmental Disabilities* 16 (May–June 1995): 179–204.

Mazzocco, Michèle M. M., K. T. Devlin, and S. J. McKenney. "Is It a Fact? Timed Arithmetic Performance of Children with Mathematical Learning Disabilities (MLD) Varies as a Function of How MLD is Defined." *Developmental Neuropsychology* 33, no. 3 (2008): 318–44.

Miura, Irene T., and Yukari Okamoto. "Language Supports for Mathematics Understanding and Performance." In *The Development of Arithmetic Concepts and Skills: Constructing Adaptive Expertise*, edited by Arthur J. Baroody and Ann Dowker, pp. 229–42. Mahwah, N.J.: Lawrence Erlbaum Associates, 2003.

Moyer, Margaret B., and John C. Moyer. "Ensuring That Practice Makes Perfect: Implications for Children with Learning Difficulties." *Arithmetic Teacher* 33 (September 1985): 40–42.

Myers, Ann C., and Carol A. Thornton. The Learning-Disabled Child—Learning the Basic Facts. *Arithmetic Teacher* 25 (December 1977): 46–50.

National Commission on Excellence in Education. *A Nation at Risk*. Washington, D.C.: U.S. Government Printing Office, 1983.

National Commission on Mathematics and Science Teaching for the 21st Century. *Before It's Too Late: A Report to the Nation from the National Commission on Mathematics and Science Teaching for the 21st Century*. Jessup, Md.: Education Publications Center, U.S. Department of Education, 2000.

National Council of Teachers of Mathematics (NCTM). *Curriculum and Evaluation Standards for School Mathematics*. Reston, Va.: NCTM, 1989.

———. *Professional Standards for Teaching Mathematics*. Reston, Va.: NCTM, 1991.

———. *Principles and Standards for School Mathematics*. Reston, Va.: NCTM, 2000.

———. *Curriculum Focal Points for Prekindergarten through Grade 8 Mathematics: A Quest for Coherence*. Reston, Va.: NCTM, 2006.

Piaget, Jean. "Development and Learning." In *Piaget Rediscovered*, edited by Richard E. Ripple and Verne N. Rockcastle, pp. 7–20. Ithaca, N.Y.: Cornell University, 1964.

Pressley, Michael, Fiona Goodchild, Joan Fleet, Richard Zajchowski, and Ellis D. Evans. "The Challenges of Classroom Strategy Instruction." *Elementary School Journal* 89 (January 1989): 301–42.

Rathmell, Edward C. "Using Thinking Strategies to Teach Basic Facts." In *Developing Computational Skills*, 1978 Yearbook of the National Council of Teachers of Mathematics (NCTM), edited by Marilyn N. Suydam and Robert E. Reys, pp. 13–50. Reston, Va.: NCTM, 1978.

Resnick, Lauren B. "Syntax and Semantics in Learning to Subtract." In *Addition and Subtraction: A Cognitive Perspective*, edited by Thomas P. Carpenter, James M. Moser, and Thomas A. Romberg, pp. 136–55. Hillsdale, N.J.: Lawrence Erlbaum Associates, 1982.

———. "Treating Mathematics as an Ill-Structured Discipline." In *The Teaching and Assessing of Mathematical Problem Solving*, edited by Randall I. Charles and Edward A. Silver, pp. 32–60. Reston, Va.: National Council of Teachers of Mathematics, 1989.

Reyes, Laurie H. "Affective Variables and Mathematics Education." *Elementary School Journal* 84 (May 1984): 558–81.

Rickard, Timothy C., and Lyle E. Bourne Jr. "Some Tests of an Identical Elements Model of Basic Arithmetic Skills." *Journal of Experimental Psychology: Learning, Memory, and Cognition* 22 (September 1996): 1281–95.

Rickard, Timothy C., Alice F. Healy, and Lyle E. Bourne Jr. "On the Cognitive Structure of Basic Arithmetic Skills: Operation, Order, and Symbol Transfer Effects." *Journal of Experimental Psychology: Learning, Memory, and Cognition* 20 (September 1994): 1139–53.

Rivera, Diane M. "Third Invited Response: Examining Mathematics Reform and the Implications for Students with Mathematics Disabilities." *Remedial and Special Education* 14 (November–December 1993): 24–27.

Rourke, Byron P., and John D. Strang. "Subtypes of Reading and Arithmetic Disabilities: A Neuropsychological Analysis." In *Developmental Neuropsychiatry*, edited by Michael Rutter, pp. 473–88. New York: Guilford Press, 1983.

Scheid, Karen. *Cognitive-Based Methods for Teaching Mathematics to Students with Learning Problems.* Columbus, Ohio: UNC Resources, 1990.

Schoenfeld, Alan H. *Mathematical Problem Solving.* New York: Academic Press, 1985.

———. "Learning to Think Mathematically: Problem Solving, Metacognition, and Sense Making in Mathematics." In *Handbook of Research on Mathematics Teaching and Learning*, edited by Douglas A. Grouws, pp. 334–70. New York: Macmillan, 1992.

Seo, K.-H., and Herbert P. Ginsburg. "'You've Got to Carefully Read the Math Sentence . . .': Classroom Context and Children's Interpretations of the Equals Sign." In *The Development of Arithmetic Concepts and Skills: Constructing Adaptive Expertise*, edited by Arthur J. Baroody and Ann Dowker, pp. 91–104. Mahwah, N.J.: Lawrence Erlbaum Associates, 2003.

Siegler, Robert, and Eric Jenkins. *How Children Discover New Strategies.* Hillsdale, N.J.: Lawrence Erlbaum Associates, 1989.

Sokol, Scott M., Michael McCloskey, Neal J. Cohen, and Donna Aliminosa. "Cognitive Representations and Processes in Arithmetic: Inferences from the Performance of Brain-Damaged Subjects." *Journal of Experimental Psychology: Learning, Memory, and Cognition* 17 (May 1991): 355–76.

Strauss, Alfred A., and Newell C. Kephart. *Psychopathology and Education of the Brain-Injured Child. Vol. II. Progress in Theory and Clinic.* New York: Grune and Stratton, 1955.

Swanson, H. Lee, and John Cooney. "Strategy Transformation in Learning Disabled and Nondisabled Students." *Learning Disability Quarterly* 8 (Summer 1985): 221–30.

Swanson, H. Lee, and Barbara Rhine. "Strategy Transformations in Learning Disabled Children's Math Performance: Clues to the Development of Expertise." *Journal of Learning Disabilities* 18 (December 1985): 596–603.

Thorndike, Edward L. *The Psychology of Arithmetic.* New York: Macmillan, 1922.

Thornton, Carol A. "Emphasizing Thinking Strategies in Basic Fact Instruction." *Journal for Research in Mathematics Education* 9 (May 1978): 214–27.

Thornton, Carol A., and Nancy S. Bley. "Problem Solving: Help in the Right Direction for LD Students." *Arithmetic Teacher* 29 (February 1982): 26–41.

Thornton, Carol A., and Nancy S. Bley, eds. *Windows of Opportunity: Mathematics for Students with Special Needs.* Reston, Va.: National Council of Teachers of Mathematics, 1994.

Thornton, Carol A., Graham A. Jones, and Margaret A. Toohey. "A Multisensory Approach to Thinking Strategies for Remedial Instruction in Basic Addition Facts." *Journal for Research in Mathematics Education* 14 (May 1983): 198–203.

Thornton, Carol A., and Paula J. Smith. "Action Research: Strategies for Learning Subtraction Facts." *Arithmetic Teacher* 35 (April 1988): 8–12.

Thornton, Carol A., and Margaret A. Toohey. "Basic Math Facts: Guidelines for Teaching and Learning." *Learning Disabilities Focus* 1 (Fall 1985): 44–57.

Trafton, Paul R., and Alison S. Claus. "A Changing Curriculum for a Changing Age." In *Windows of Opportunity: Mathematics for Students with Special Needs*, edited by Carol A. Thornton and Nancy S. Bley, pp. 19–39. Reston, Va.: National Council of Teachers of Mathematics, 1994.

U.S. Department of Education. *Foundations for Success: The Final Report of the National Mathematics Advisory Panel.* Washington, D.C.: U.S. Government Printing Office, 2008.

van Erp, Jos W. M., and Lous Heshusius. "Action Psychology: Learning as the Interiorization of Action in Early Instruction of Mathematically Disabled Learners." *Journal of Learning Disabilities* 19 (May 1986): 274–79.

Wertheimer, Max. *Productive Thinking.* New York: Harper and Row, 1959.

Wilson, Suzanne M. *California Dreaming: Reforming Mathematics Education.* New Haven, Conn.: Yale University Press, 2003.

CHAPTER 3

Instruction: Yesterday I Learned to Add; Today I Forgot

Jeffrey Shih,
William R. Speer,
and Beatrice C. Babbitt

A primary consideration in the effective teaching of mathematics is that students bring a wide range of abilities and learning approaches. Teachers must recognize, reveal, and address these differences among learners. We frame this chapter according to the National Council of Teachers of Mathematics Teaching Principle (NCTM 2000):

Effective Mathematics Teaching Requires Understanding What Students Know and Need to Learn and Then Challenging and Supporting Them to Learn It Well

To truly embrace the vision and promise of the *Principles and Standards for School Mathematics* (NCTM 2000), we must revitalize mathematics programs and rethink teaching and learning for the benefit of all students (Cawley and Reines 1996; Miller and Mercer 1997). The following example shows the complexities of teaching and learning for the benefit of *all* students:

> Ms. Sanchez knows that even with the best of planning, things do not always turn out as expected. However, today's results were particularly discouraging. Ms. Sanchez had been excited about the mathematics activity that she had planned for her fourth-grade mathematics class. It was a relatively simple activity requiring students to explore the relationships among proper fractions by using fraction bars, a ruler, and possibly some measuring cups. Students were supposed to work in groups and determine the order of magnitude of a given set of fractions. They were also encouraged to add to the set of fractions and incorporate these new fractions in the sequence. Students were to record any work and describe the accompanying experiments that supported their conclusions. Ms. Sanchez believed that she had accommodated the lesson to meet the special needs of her students: she had incorporated several concrete materials and ensured that each group included a natural leader and a student with good writing skills. They had spent the last few days learning how to represent fractions. Surely the students had all they needed to succeed with today's activity. As Ms. Sanchez moved from group to group, however, she saw several problems

developing. Sara had taken over group one and was busily marking the position of the fractions on the ruler while the other students made designs with the fraction bars. Jeff decided to place fraction bars on top of one another to determine size, but since others could not see what he was doing they busied themselves scooping rice with the measuring cups. Lori and Kyle were intently arguing about the relationship between 2/3 and 4/5 while their groupmates created fences with the fraction bars. What Ms. Sanchez had designed to be an interesting exploratory activity had turned into an objective-driven learning activity for only a few. What went wrong, and why? How could Ms. Sanchez have more effectively designed and carried out the lesson for this, or any, diverse population?

When one considers differences among students, the crucial questions include the following:

◆ How should teachers motivate and encourage every student to effectively explore and continue studying mathematics?

◆ How should schools and school districts develop and correct policies, programs, and practices that may contribute or lead to mathematics avoidance?

◆ How should mathematics educators make individual and collective commitments to eliminate barriers to studying mathematics?

◆ How should educators explore and implement effective means of convincing students and stakeholders of the importance of mathematics as a field of study?

Consistent with this book's purpose, this chapter will focus on challenging and supporting all students in learning mathematics—particularly children who often struggle to understand mathematical concepts, tend to expend great effort to calculate accurately, and seem to lack persistence in problem-solving situations.

Reform documents rarely give much guidance on modifying circumstances for at-risk students or those with a diagnosed learning disability in mathematics. Researchers in mathematics education often focus on anecdotal accounts of the effects of reform-based pedagogy and curricula on low achievers (Fennema et al. 1993). Some researchers seem to imply that reform-based mathematics pedagogy and materials are effective for all students, without a need for any special adaptations to curriculum, instructional techniques, or classroom organization (Resnick et al. 1991).

Some researchers in special education express doubt that proposed methods and materials associated with reform mathematics are appropriate for students with learning disabilities or those at risk (Carnine, Dixon, and Silbert 1998; Carnine, Jones, and Dixon 1994; Hofmeister 1993). For example, special educators have long recommended using a clear set of procedures to reduce ambiguity when teaching mathematics (Carnine, Jones, and Dixon 1994). Believing that multiple approaches to solving problems can lead to confusion, these researchers view alternative strategies and invented algorithms, a common approach in reform-based mathematics instruction, as problematic for low achievers. These researchers see one simple set of rules as the best approach to teaching these students.

Research on attempts to achieve inclusion for special education students, particularly students with learning disabilities, also suggests that general education teachers have a hard time accommodating such students' needs (Baker and Zigmond 1990; Schumm et al. 1995; Scruggs and Mastropieri 1996). Researchers have typically tried inclusion in settings where general education teachers used traditional pedagogy and curricular materials. Two dramatically different interpretations of inclusion have emerged. One view is that traditional pedagogy is, in some sense, responsible for the difficulties that low achievers experience and that, with curricula and pedagogy that emphasize levels of both content and pedagogical reform, many students who formerly struggled in traditional mathematics instruction will thrive. A contrasting view is that students who have difficulties in a more traditional mathematics setting will have even greater problems with the advanced topics and problem-solving activities in a classroom emphasizing a more reform-based approach. We need additional classroom-based research to clarify and perhaps end such debates. The lack of research, coupled with the concerns of the special education community, highlights the need for studies on how reform mathematics instruction affects low achievers.

Bottge (2001) posed the *key–lock model* of teaching mathematics to help guide efforts to improve learning of students with disabilities (fig. 3.1). The model is based on theories of cognition; emphasizes the Equity Principle (NCTM 2000, p. 12); and considers learner, contextual, and task variables essential to adequately describe teaching and learning mathematics. For significant learning to occur, the model proposes, six teeth of the instruction key (meaningful, explicit, informal, [de]situational, social, and teacher specific) must each fit a pin of the learning lock (engagement, foundations, intuitions, transfer, cultural supports, and student specific).

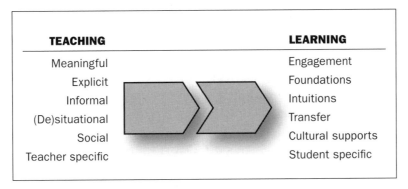

Fig. 3.1. The Bottge key–lock model

The model represents teaching and learning and is not a prescriptive guide. For example, the model leaves to the teacher finding meaningful learning experiences to engage students and ways to teach foundation skills explicitly. The teacher must also use informal methods to encourage students to use the intuitions they bring to the classroom and to (de)situate learning experiences, which students will recognize and transfer to problems in future contexts. And because students often solve authentic problems in groups, teachers should encourage students to work together in social contexts where they can find support for their ideas. Somewhat nonscientifically, the model also acknowledges teacher and learner interpersonal factors that either contribute to or work against higher achievement.

Even so, the Bottge key–lock model gives us a useful metaphor to explore the parameters of teaching and learning in a special-needs construct.

Effective Teaching Requires Knowing and Understanding Mathematics, Students as Learners, and Pedagogical Strategies

Knowing and understanding mathematics is a challenge for far too many elementary and special education teachers. Love of children and interest in developing reading and writing skills may be of greater interest to these teachers. In short, they are far more comfortable teaching subjects other than mathematics. Moreover, some middle school and high school teachers holding initial certification in fields other than mathematics find themselves assigned to mathematics classrooms. Such teachers may find themselves beginning to teach with inadequate content knowledge and skills in mathematics. Teachers may find fewer resources for students with special needs in areas such as algebra and geometry (Babbitt 2006; Witzel, Mercer, and Miller 2003). When they must accommodate students with learning difficulties, teachers may feel more anxiety from lack of awareness or knowledge of student needs and appropriate instructional strategies. Despite these limitations, most teachers want to, and strive to, teach effectively; the challenge is how to meet these diverse needs. This chapter discusses instructional strategies that will help teachers support mathematics learning for students with learning difficulties. These strategies, combined with opportunities for increased teacher content knowledge, are key to meeting the NCTM goals of mathematics for all.

Understanding learners with special needs suggests recognizing that students do not all learn in the same way (Badian 1999; Fox 1998; Keeler and Swanson 2001). Although students across achievement levels often have difficulty understanding concepts and skills, these difficulties often persist longer with special-needs children (Miller and Mercer 1997; Patton et al. 1997; Shalev et al. 1998). Also, these difficulties often resurface when teachers introduce new, more complex ideas and procedures. For example, subtracting from zero can be a problem for many children. Although all students might benefit from place value modeling, students with learning difficulties will initially need extensive place value modeling to understand this concept. They may also require more work on judging the reasonableness of an answer to catch their own errors, as well as systematic practice to overcome the common tendency to write "20 – 6 = 26." Such students will often revert to this error when attempting long division.

One approach to meeting the diverse needs of students in today's classroom is incorporating multiple representations of mathematical ideas. Doing so increases the probability that teachers will reach every student through an efficient and effective personal learning style. Different students benefit from hearing "it," seeing "it," saying "it," touching "it," manipulating "it," writing "it," or drawing "it." Most students benefit from experiencing a mathematical concept in several of these ways. Most students also expand personal understanding of a concept by seeing it represented with different materials or described with different examples. Before using materials and representations, teachers must carefully assess learning styles—both how the students take in the information and how they output it (receptive and expressive modalities). Using multiple representations does not mean bombarding them with these at the same time. Teachers should exercise caution in the number of presentation modalities used, particularly in the early stages of learning. A mathematical concept's seemingly unlimited number of different representations may simply overwhelm many students with learning difficulties. They may perceive each representation as a new concept and fail to see the commonality and relationships among the varied representations. Multiple stimuli and multiple representations could obscure the mathematical concept they were designed to clarify.

Consider the ways to represent decimals with dollars, dimes, and pennies; with place value blocks; and with a metric ruler. Each representation has its advantages: real-life, familiar materials used in new ways. Each representation also has its disadvantages: money relationships, such as relative value, are not obvious from each of the materials; the thousands cube (or, more commonly, the hundreds flat) now becomes the unit block; the metric system may be unfamiliar to the students and thus not serve as a good model for decimals. Each system requires a solid understanding of the unit and the relationship among the objects within the system. Students with learning difficulties often have problems keeping within-system relationships in mind without the added difficulty of comparing across representational systems.

From a cognitive perspective, however, teachers must represent mathematical ideas in multiple ways to expand students' understanding of core ideas and to help them see connections among these ideas. The challenge to the teacher is not only to help students link ideas and concepts superficially but also to establish an ownership of the relationship. Students with learning difficulties rarely learn from seeing or hearing only. In a cooperative group setting, someone else's having discovered a relationship is typically not enough to ensure understanding by students with learning difficulties—nor is having one person in the group summarize the discovery. All students must personalize the learning by demonstrating it with many examples. All students must also be able to describe the newly found relationship in their own words. They must be able to demonstrate understanding in some way that is meaningful to them, perhaps by using drawings or models, orally, or in writing. Moreover, few students, particularly those with learning difficulties, can retain and use a new connection after one encounter. They will need frequent experiences with the new connection on later days for it to become part of their conceptual repertoire.

The *Principles and Standards'* Content Standards (NCTM 2000) and the *Curriculum Focal Points for Prekindergarten through Grade 8 Mathematics: A Quest for Coherence* (NCTM 2006) and, more recently, the *Common Core State Standards* (Council of Chief State School Officers [CCSSO] 2010) all stress the need to understand numbers and the relationships among them, to comprehend the meaning of operations, and to compute fluently and make reasonable estimates. The research literature on instruction often addresses understanding as concept instruction, whereas it often addresses computation under basic skill instruction. Extensive special-needs research supports systematic instruction in basic skills that incorporates appropriate modeling, sustained practice, and planned reviews (Hudson and Miller 2006; Fuchs and Fuchs 2001; Mastropieri, Scruggs, and Shiah 1991). Special-needs students experience enhanced mathematical performance when teachers explicitly model computational procedures (Miller and Mercer 1997), model thought processes used during the procedure (Montague 1997), offer sustained practice on a new skill (Fuchs and Fuchs 2001), and schedule planned reviews that slowly integrate new and old skills while leaving no skill unpracticed for long periods (Carnine 1997).

Although most teachers explicitly model new procedures to the entire class, they may find it necessary to redo this modeling with a small group of students with learning difficulties to make certain that these students comprehend each step as modeled. Materials that help students organize their work are particularly helpful for students who have difficulty organizing problems on a page and keeping numbers aligned. Examples include using lined paper rotated through ninety degrees to align numerals, using black construction paper "windows" to block out the excess stimuli on a page, and simply folding back the page to expose less material at one time. Students with language learning difficulties will also need particular help learning the self-talk that can guide personal movement through the process. Practice on the skill will need to continue over many days, rather than only a few days, with short daily reviews supporting this development. Systematic integration of ever more difficult problems will help students gain the desired competence and fluency.

Conceptual understanding is an important component of mathematical proficiency but has received less attention in the special-needs literature (Baroody and Ginsburg 1991; Ginsburg 1997). However, Miller and Mercer (1997) established the importance of concrete experiences, pictorial modeling, and gradual transitions to abstract thinking. Students with learning difficulties in mathematics often profit from beginning instruction in which they use their own bodies or concrete materials to illustrate mathematical ideas. Students learn a counting sequence as they count themselves while lining up for recess. Students distribute real apples to classmates to illustrate the sharing model of division. Students determine equivalent fractions by using objects that have meaning in their everyday world as fraction manipulatives.

Pictorial modeling helps student thinking become slightly more abstract. Ready-made pictures or drawn pictures can help illustrate a mathematical idea such as proportion. Early on in their use, pictures also should be able to be manipulated since many special-needs students have difficulty imagining a change in the pictured situation. For example, some pictured images that are supposed to convey movement, as in common set-separation pictures for subtraction, may not be easily interpreted as subtraction by the attempt to stagnantly show birds flying away from other birds or koalas climbing down trees or frogs hopping off lily pads.

As instruction moves to the symbolic level, teachers must form a link between the prior concrete and pictorial representations and the now-new abstract representations (Miller and Mercer 1997). Teachers must make this link, or transition, explicit so that the student knows exactly how symbols represent each element of the concrete or pictorial representation. Without this crucial link, students who understand at the concrete and pictorial level will often revert to mindless, and thus potentially meaningless, actions when using symbols. Butler and colleagues (2003) showed the effectiveness of this *concrete–representational–abstract* teaching approach in effectively teaching fraction concepts and operations.

Problem solving

Problem solving has been an area of particular concern for students with learning difficulties (Bottge 2001; Montague, Applegate, and Marquard 1993). Nearly every aspect of problem solving appears to pose some form of difficulty for students with special needs. If the teacher describes a problem to the class, some students may not be able to hold it in mind long enough to begin to solve it. However, if the problem is written, some students do not attempt to read it for comprehension but rather try to solve it by simply operating on the numbers. Students may not read the problem because of problems with decoding or reading comprehension. If the students do read the problem for comprehension, they may have difficulty picturing the problem situation. Even if students understand the problem situation, they may not make the leap to a solution strategy. In such instances, all strategies seem to hold equal value and promise. Strategies that require computation introduce the prospect of computation errors into problem solving. If students do arrive at a solution, they may rarely check to make sure that an answer is reasonable and accurate. Despite these challenges, students with learning difficulties improve problem-solving performance with appropriate instruction (Montague 1997).

Research has suggested a variety of problem-solving approaches for students with learning difficulties. Many approaches suggest following a series of problem-solving steps (Hutchinson 1993). Most of these approaches work well with traditional word problems but are less helpful with the nontraditional problems more common in standards-based classrooms.

Some have proposed anchored instruction as an effective alternative to step-by-step problem-solving strategies. Anchored instruction presents real-life situations as contexts (the anchors). We might consider problem solving within a traditional school store setting to be anchored instruction. Current versions of anchored instruction often use video vignettes, through which any real-life experience with embedded mathematical data might serve as an anchor for problem-solving instruction. For example, some videos focus on discovering measurements embedded in such adventures as fighting a fire or on interpreting medical measurements that emergency teams use. Whereas a typical problem that a teacher or textbook poses depends on the experience and imagination of the students, the videos compensate for different experience levels by giving all students a common picture of the problem situation or setting. The problem situation is reality based and complex enough to incorporate data for several problems. Students can refer to the video at any time to check information. The Cognitive Group at Vanderbilt University reported increased persistence and growth in problem-solving skills when using anchored instruction for students with learning difficulties (Cognition and Technology Group at Vanderbilt 1990).

Problem-solving environments such as those in the *Principles and Standards* are complex, often posing nontraditional problems. These guidelines respect and even encourage multiple approaches to problem solving. Language plays an important role in communicating the problem-solving approaches that students use. Students with learning difficulties may struggle within this environment, and teachers may wonder how to help them find their way.

To assess student needs, teachers must reflect on what happens during these complex problem-solving interchanges. Teachers also need to be open to observation by and feedback from colleagues to help understand the challenges to learning in their classroom (Friend and Cook 1996). A colleague–observer might be able to collect data to determine whether the teacher has structured open-ended questions so that struggling students have an opportunity to lay out the path to a solution. For example, an observer might note whether a teacher moves on after the top five students respond or instead offers all students a chance to respond in ways appropriate to their skills and abilities. Although a third party as an external observer will be beneficial to help make sense of the complexity in the standards-based classroom, entering into this professional growth partnership nevertheless requires collegiality and professionalism from both parties.

Dispositions

Principles and Standards describes the desired dispositions of children learning mathematics (NCTM 2000, p. 54). These dispositions encompass the desire that all students become autonomous learners who define personal goals, monitor personal progress, display confidence in their ability to solve a problem, show eagerness to figure things out on their own, are flexible in exploring mathematical ideas, and are willing to persevere.

Children with learning difficulties in mathematics rarely display these dispositions and in fact often display opposite behaviors. Such children are often adult dependent and display characteristics of learned helplessness (Halmhuber and Paris 1993). They may lack confidence in their ability to perform in mathematics, enthusiasm for mathematical activities, and flexibility or perseverance (Deshler, Ellis, and Lenz 1996). However, as *Principles and Standards* states, "Effective teachers recognize that the decisions they make shape students' mathematical dispositions and can create rich settings for learning" (NCTM 2000, p. 18).

Two specific teacher behaviors will begin to support positive changes in student dispositions. First, students must feel that they are in an environment where taking risks is safe. Students with learning difficulties—and most students, for that matter—stop taking risks when their approaches are wrong and when others ignore or ridicule them for being wrong. The teacher's challenge is to help students see that their path is valued and protected, as well as to guide them along that path. . . . Sometimes the path they have chosen may seem wrong to us but actually is one that may take us to a solution if we encourage and nurture the persistence that these students need. Students need to be right more often so that they will try again, and they need help to find the kernel of a useful idea in an incorrect response that can lead to eventual understanding. Second, since standards-based classrooms use many cooperative groups and peer groups, all students must learn these nurturing approaches as well. Even kind peers will tend to ignore a peer's wrong response and propose a more appropriate solution. Learning theory suggests that punishment, ridicule, or consistent ignoring will extinguish a behavior. Here, such treatment might extinguish the desirable behavior of contributing strategies for problem solving unless teachers and peers alter interactions with students.

Effective Teaching Requires a Challenging and Supportive Classroom Learning Environment

Expanding the range of pedagogical strategies is one important consideration for teaching children with special needs. Teachers must pay greater attention to length and organization of tasks than is the norm for a particular age level. They must also take into account the impact of oral and written language difficulties as students demonstrate what they have learned. If teachers make clear the primary purpose for each mathematical activity, they will be better able to adjust requirements in supportive domains. That is, if the purpose of a given lesson centers on problem solving, then teachers might give aided support of computation for those students who still have a hard time with the required computation.

Challenging students without overwhelming them is an important task for teachers. Teachers need to individualize challenges but to consider them systematically so that all students move forward in the ability to handle ever more complex mathematical tasks. A key to challenging students with learning difficulties is realizing that they can often think about the same challenging problems as classmates but that they may need more support in organizing responses. For example, one activity for all students early on in data collection, organization, and analysis might be to develop several reporting systems that student groups can then use as desired. Students could then compare various data organization systems as far as usefulness in organizing and reporting data. Many students with learning difficulties will find that these organization systems help them approach a problem systematically without getting lost in the details. They will also help teachers and peers give specific feedback about which parts of the task remain to be completed.

Active engagement

Challenging environments require active engagement from all students during whole-group discussion, as well as small-group and paired activities. However, Baxter, Woodward, and Olson (2001) present a sobering picture of low achievers' involvement in whole-class discussions and small-group activities. In the classrooms that these authors observed, low achievers and students with special needs were generally silent during whole-group discussions. When such students did speak, they often gave one-word answers. During the class discussion in which other students might be actively engaged, the low achievers were often perceived not to be listening but instead were involved in off-task behaviors such as staring out the window, playing with a small toy, writing on paper, or arranging materials in their desk. Low achievers were much more involved in small-group work, but they often took nonmathematical roles such as organizing the materials or copying a partner's work. Pairs of low achievers had a difficult time starting the task and needed help from the teacher or aide to reread the directions, start the task, and work through several problems.

    Some teachers have found ways to engage students with learning difficulties in meaningful mathematical tasks (Baxter, Woodward, and Olson 2001). Low-achieving students were engaged when the teacher first used the children's own arms to model parallel lines, then used yarn with children as endpoints to create parallel lines, and finally moved students to using a geoboard to model parallel and intersecting lines. The teacher emphasized both the mathematical vocabulary and the conceptual ideas during the lesson. Baxter, Woodward, and Olson speculated that the teacher's use of multiple representations before using the traditional geoboard made this lesson more effective. A special education teacher might recognize that this teacher first had the children experience the concept with their own bodies, then they externalized the concept by using yarn with themselves as endpoints, and finally they transferred this learning to objects totally outside themselves: rubber bands and pegs on the geoboard. Even though these children were third graders, they learned best when engaged in activities that were similar to those used with younger children. They could understand age-appropriate mathematical concepts by using learning approaches that others might consider developmentally delayed.

Adapting instruction

Several effective methods exist to adapt instruction for students with learning difficulties. Those described here include scaffolding; time adjustments; homework adaptations; and using technology in instruction, including assistive technology. We encourage you to seek out other techniques in the professional literature that this and other chapters cite.

Scaffolding

Scaffolding refers to the supports that teachers give during the early stages of learning a concept or skill. Scaffolds furnish temporary support as a student learns a new skill. The idea is often linked to Vygotsky's (1962) zone of proximal development, wherein one chooses instructional levels that are neither too easy nor too difficult for the child. Vygotsky views the social interaction of the teacher and student within this zone of proximal development as crucial for effective learning. In mathematics, scaffolds might include rephrasing a problem to make it more understandable, modeling a problem solution, thinking aloud by the teacher to show the thought process required in the problem solution, and questioning to help guide a student through problem solving. Scaffolds assist all students in developing the metacognitive skills that support problem solving. Scaffolds also often supply the vocabulary to discuss the learning experience, which is particularly important for students with language deficits.

Time needed to learn mathematics

Instructional time is valuable and limited, yet students with disabilities may require more exposure when learning a mathematical concept, more practice mastering a computational procedure, and more experience with each aspect of problem solving than other classmates. Adjusting for variances in needed learning time is essential in every classroom.

If teachers ignore this issue, many students with special learning needs will not learn any mathematical concept or skill well; the class will always seem to move on to a new concept or skill before the student with special needs is ready. However, no easy solutions exist for this scenario. Our colleagues researching reading are looking at offering intense instruction (one-on-one tutoring) as one means to overcome the time limitation (Richek et al. 2002). Parents, older students, and senior volunteers are some of the resources whom teachers might enlist to better use available time and extend the actively engaged learning time for students with learning difficulties. One thing is clear: we will get improved results in mathematical performance only if we change how teachers use time for students with learning difficulties.

Time children will be able to stay on task

Even in the best of instructional worlds, where teachers fully use available instructional time, time on task will remain an issue for many students—those with and without learning difficulties. Many students with learning difficulties have trouble focusing on one task for an extended period. This situation is actually a relatively easy area to adapt in a mathematics classroom. Teachers can break instruction and assignments into segments that better fit the student's ability to remain on task. They can mix group and individual activities to vary input and output demands. Educators can use activities with frequent feedback to maintain student attention; technology-assisted instruction often incorporates this feature, but self-correcting materials or interactive mathematical games may also incorporate feedback mechanisms. Teachers can incorporate activities that use manipulatives into paper-and-pencil activities. They can break up longer assignments by including required checkpoints. Instructors can ask problem-solving groups to pause every ten minutes to review accomplishments and reiterate agreed goals. Teachers may be surprised to learn that students with attention span difficulties often find more success in a classroom that moves faster than the norm, with lots of interaction and frequent activity changes. Such a classroom must be well organized, with smooth transitions between activities so that time gained during instruction is not lost in transitions.

Homework

Homework can extend student opportunities to think about mathematical concepts, practice for automaticity, and problem solve using common household items. Homework assignments are most helpful when teachers carefully plan them and they have direct meaning to the students (Paulu 1995).

Homework can maintain skill for and give practice to students. Students should be proficient enough with the skills and the tasks before practicing without supervision to prevent automatizing incorrect procedures (Kilpatrick, Swafford, and Findell 2001, p. 480). Particularly for students with learning difficulties, teachers should ensure that homework is realistic in length and difficulty. In assigning homework, teachers should give clear directions and offer meaningful activities closely linked to classroom instruction. To ensure that homework will have instructional value—and to convey the important role that homework plays—teachers must check and review their students' completed assignments.

Homework can also help maintain communication between parents and teachers. Homework assignments not only let parents become familiar with what their children are studying but also allow extending the actively engaged learning time for students with learning difficulties.

Role of technology in instruction

Access to and familiarity with technology is an issue for special-needs teachers as well as for special-needs students. Suffice it to say at this point that even simple calculators can be used during concept development, to do or check computations, and to support problem-solving activity. Graphing calculators can support work with functions, graphs, and data analysis, as well as other mathematical concepts.

Although most students can use calculators effectively, some students with disabilities will make calculator errors that are similar to the computation errors they make in other settings. That is, they may enter data without checking the entry; they may enter data and operations in the wrong sequence; and they may fail to correctly interpret output, particularly decimals. Students may also fail to understand the purpose and use of various designated function keys such as memory keys and hence will not have access to some of the efficiency benefits of calculator use.

The advent of digital technology has extended the basic idea of manipulatives to computer-based, or virtual, manipulatives. A wide range of virtual manipulatives, such as those available through the National Library of Virtual Manipulatives (nlvm.usu.edu/en/nav/siteinfo.html), can support instruction and learning of many mathematics topics. The dynamic and interactive features of virtual manipulatives make them ideal for illustrating many mathematics concepts. Virtual manipulatives can supply the visual and graphic depiction of problems that is effective with students with difficulties in mathematics (Gersten et al. 2009). While research into the use of emerging technologies continues, there are some variables to consider when using, or measuring the effects of, virtual manipulatives. Research design, sampling characteristics, and type of manipulative used may each influence student achievement. For example, studies show evidence of increased achievement when classroom teachers believed virtual manipulatives fit in with the natural flow of the curriculum. Other variables that may influence the effectiveness of using virtual or physical manipulatives include previous experience with computers, grade level, mathematical topic, treatment length, student attitudes toward mathematics, and computer-to-student ratio (Heddens, Speer, and Brahier 2009).

Assistive technologies can help students gain access to and benefit from their educational programs. Talking calculators, Braille keyboards, captioned videos, and alternative computer input devices are just a few of the assistive technologies that teachers can incorporate into their instruction. The Georgia Project for Assistive Technology (www.gpat.org) offers an extensive list of devices for students struggling with mathematics. It usually takes time for both teachers and students to become comfortable using assistive technologies before they are likely to be incorporated effectively into mathematics instruction.

Adapting materials

Teachers can adapt many features of materials, models, and manipulatives used to explore mathematics to make these implements more user friendly for students with learning difficulties. Clarity is an important feature to incorporate into the materials and models we use. Teachers should use white space liberally to separate problems and ideas. A consistent format will benefit the student in using the material. The reading level of materials may exceed that of the students using the materials. Teachers may need to rewrite or orally reword directions, explanations, and word problems or to arrange for a peer or aide to read the material to the student.

The focus of attention on any given material or model should be the underlying mathematical concept or skill. Educators should replace graphics only remotely related to the skill with relevant graphics that clearly depict a concept. An example would be to *not* use a picture of a collection of objects, say, five books, with an *X* drawn on two of them to represent subtraction. The resulting image would *still* have five books, just three without *X*s and two with *X*s—which does not represent an image of "5 – 2" by take-away, comparison, or missing addend (fig. 3.2).

Many students with disabilities have difficulty with visual–spatial organization, which interferes with their ability to organize work so they can follow their own reasoning or compute accurately. These students often profit from materials that help them organize

Fig. 3.2. X-ing out images does not truly represent subtraction.

their work on a page. Teachers can use many simple techniques to help children be organized and ready to learn mathematics. Simply dividing a page into four, six, or eight sections (depending on age and size of handwriting) can guide a student in arranging problems on the page. Rotating lined paper ninety degrees will give students columns for multidigit operations. Using appropriate-sized graph paper is another technique to align columns. Taking any operation and furnishing a skeleton framework for the steps of an algorithmic computation will help many students. Increasing the font size and the room to work problems will help many students whose writing is large. Giving cues on worksheets is often helpful in the beginning stages of learning an algorithm; such cues include an arrow to indicate where to begin or a box to contain the number regrouped. A recording worksheet to accompany a problem-solving activity might be separated into stages. Initial worksheets to accompany manipulative activities would offer enough space for students to outline the actual manipulative when drawing it. For manipulative-based activities, supplying mats as a work surface and containers for extra manipulatives will help students stay organized. Teachers should always be alert to student disorganization that interferes with clear thinking and accuracy.

Adapting the learning environment

This section covers various approaches to adapting the learning environment for students.

Whole-class instruction

Whole-class instruction presents major challenges to students with special needs. The pace required to keep the attention of a large group rarely matches the needs of a struggling learner. Although excellent teachers will model mathematical procedures and material use, many struggling students will not understand this modeling from a distance. During class discussions, students may not follow the shifting focus of attention from teacher to various students throughout the room. Hence, they may miss or misunderstand key elements of the problem-solving arguments that their classmates make. Many struggling students will also have difficulty maintaining attention throughout whole-class instruction.

A variety of models for whole-group instruction exist. Explicit instruction is one approach to whole-group instruction that research has validated with students with diverse learning needs (Greenword, Arreaga-Mayer, and Carta 1994). Clear and direct presentation of concepts and skills marks this approach (Miller 2002), in which teachers establish clear goals, monitor student performance throughout the lesson, and give feedback. Typical explicit instruction incorporates an advance organizer, a description and demonstration of the concept or skill, guided practice, independent practice, and a postorganizer. The advance organizer links new learning to already-acquired knowledge. Both describing and demonstrating are part of the lesson body. Teachers describe concepts and skills and illustrate them with manipulatives, pictures, or diagrams as appropriate. Teacher support is an important element of guided practice. Students try out the new procedure or skill with teacher monitoring and feedback. If students are struggling excessively, the teacher can pull the entire group back into a clarifying description and demonstration mode. Over time, the teacher offers less and less support as students become independent performers. Independent practice is directly linked to the instructional goal and uses the same approach to problem solving used during the lesson. For example, if a lesson used place value blocks in solving addition with regrouping problems, then independent practice would use similar materials. Teachers and texts often violate this instructional guideline by asking students to solve abstract problems on their own after participating in a guided lesson that used concrete materials. Most students with learning problems cannot make the transition to independence and abstraction simultaneously and without further instruction in abstract-only procedures.

In contrast to explicit instruction, many whole-group lessons, consistent with the NCTM *Principles and Standards*, incorporate problem-solving activities as the starting point to concept and skill development. These guidelines encourage multiple solution strategies. Sharing these varied solutions is an important part of the lesson. The strength of this approach is that students at various levels of mathematical development can solve meaningful problems in ways consistent with their concept and skill development. This method promotes connections between concepts and skills as students share varied approaches to problem solutions. Educators using this approach often assume that students with less developed strategies will gradually adopt more sophisticated strategies as they see their classmates using them. Although this progression may be happening for some groups of previously poor-achieving students (Doty, Mercer, and Henningsen 1999; Garrison and Mora 1999), not enough research exists to support this assumption for students with disabilities (most studies of standards-based mathematics programs do not include students with disabilities in their samples). Hence, teachers may find that they must explicitly teach specific concepts, skills, and problem-solving strategies to some subgroups of students who do not acquire these during whole-group instruction with a multiple-strategy problem-solving focus.

Small group–directed instruction

Teachers often use small group–directed instruction to reteach a concept to the whole class. The teacher can move at a pace that better fits the small group of learners, can observe the responses of all students, and can rephrase or reexplain as needed. Students can respond more often, receive feedback tailored to their level of understanding, and be held responsible for all elements of the lesson. Keeping all students on task is much easier in this arrangement because of the proximity to the teacher and greater student involvement. This system often incorporates alternative teaching materials to better illustrate a concept or appeal to the small group. This forum's enhanced communication level often identifies missing background information, which the teacher can then supply. Students with special needs generally respond well to small group–directed instruction.

Special considerations in using cooperative learning

Best practices in cooperative learning must be in place for *all* students to profit. Teachers must structure groups to require full participation for *all* students. Students with special needs will not adequately profit from participation in cooperative learning groups unless teachers consider the changes that must occur in cooperative group structures. Group members must rotate roles, or the student experiencing difficulties will be relegated to a role that may not include thinking about mathematics (e.g., organizing the materials, handing them to the student who is thinking about the problem solution). All students must have a chance to use language to describe their own understanding of the problem. All students must have the opportunity to record the problem solution and the thinking of the group. If some students leap ahead to a problem solution, teachers should encourage them to go back and fill in the gaps in reasoning for other group members. Students must respect that group members will understand and solve problems at different levels of sophistication. Teachers must understand that membership in a group that presents a sophisticated problem-solving approach to the class does not, in itself, guarantee that everyone in the group will understand the approach.

Stations

Stations, which the elementary level uses more often than the secondary level, present particular opportunities and challenges for students with special needs. Computer stations may incorporate software that can offer alternative approaches to concept, skill, and problem-solving development. Technology-assisted instruction can also help students achieve accuracy and fluency in a highly interactive, feedback-intensive environment. Teachers often design activity-based stations to extend student learning. Stations should in general incorporate independent but meaningful activities that engage students in learning while the teacher is working with other groups of students. Hence, their design should incorporate features that support independent learning. Instructors should introduce new stations to students as a group. Students must know what to do to immediately begin working, what a completed task looks like, where to submit the completed task, and how to transition back to their assigned seat. They must also know what to do if they have a question. Directions must be clear and incorporate pictures or symbol prompts if the reading level of the student is a problem. Posting samples of worksheet templates or partially completed work products is helpful. Folders or envelopes to receive work should be part of the station design. Students must already have competence with the supplied tools (e.g., rulers, calculators, computer software). Teachers should assign a student helper each day to answer questions and should establish a silent request-for-help system. Teachers will probably need to show students how to effectively participate in work at stations early in the year, and they will need to review this instruction after vacations or whenever the process begins to break down. Determine how long students can work independently and create tasks that they can complete in this time frame. Gradually increase the length of tasks. Teachers might need to create a system of positive behavior supports to gradually lengthen the time that a student can be productive at an independent learning station. Increasing the interest level of the tasks may produce the same result (Baroody and Ginsburg 1991).

Effective peer tutoring

Peer tutoring is an effective instructional approach for students with disabilities (Schumaker and Deshler 1994/1995; Greenwood, Delquadri, and Carta 1997). Both the tutor and tutee benefit academically and socially from the experience. Students with learning difficulties can effectively serve in either role during same-age peer tutoring if the materials used are self-correcting and the tasks are chosen carefully. Students with disabilities can also effectively tutor younger students. Specialists have developed and validated effective classwide peer-tutoring programs for students with and without disabilities. *Together We Can! Classwide Peer Tutoring to Improve Basic Academic Skills* (CWPT; Greenwood, Delquadri, and Carta 1997) and Peer-Assisted Learning Strategies (PALS) for reading and math instruction (Fuchs, Mathes, and Fuchs 1996; Fuchs et al. 1996) are two programs that teachers have used successfully in general education classrooms with diverse student populations. Both programs train students how to be effective tutors and tutees.

Independent seatwork

Independent seatwork can take many forms but most often consists of independent practice. According to Miller (2002), independent practice serves four important functions:

1. Students can demonstrate that they have learned what they were taught.
2. Students can increase their proficiency.
3. Teachers can evaluate their teaching.
4. Students retain what they have learned.

For independent practice to achieve these goals, teachers should assign work so that accuracy levels will be high and the format of the assignment is familiar.

Opportunities to Reflect on and Refine Instructional Practice—Alone and with Others—Are Crucial

Here we discuss using the collaborative model in mathematics classes, what parents need to know about instruction, and support for teachers from the administration.

Collaborative model in mathematics classrooms

Many schools have adopted a collaborative model of service delivery for students with special needs. This model requires that people with diverse areas of expertise, such as a classroom teacher and a special education teacher, work together to find creative solutions to mutually defined problems (Langone 1998). Friend and Cook (1996) suggest that successful collaboration requires (1) mutual goals, (2) voluntary participation, (3) parity among participants, (4) shared responsibility for participation and decision making, (5) shared responsibility for outcomes, and (6) shared resources. In collaborative consultation, the special education teacher and the classroom teacher plan effective approaches to teaching the content to students with special needs. This system may emphasize ways to modify classrooms and instruction to meet student needs. Modifications include helping students become better organized, prolonging attention and concentration, improving listening skills, adapting the curriculum as needed, and helping students better manage time (Lerner 2000). Today's mathematics classroom will also require helping students to participate effectively in cooperative groups and to express mathematical reasoning. Sometimes the mathematics teacher and special education teacher will decide to coteach particular content. This arrangement will often occur if a class has many students with special needs. Both teachers supply instruction and support services when coteaching, and both should define their exact roles collaboratively. In a collaborative consultation model, the classroom teacher and special education teacher together review the mathematics instructional goals for the class and anticipate any problem areas that may require special planning. They may consider the following:

*   How will we accommodate students with reading problems?
*   Can we introduce new vocabulary that can make meaning more explicit?
*   Could we use concrete and pictorial models to illustrate a new math concept?

◆ Could we make up a flip chart of a new multistep operation so that students could refer to it when working alone?

◆ During a group problem-solving activity, could we ask students to record thought processes for later review?

◆ Could we physically enlarge problems for students who need extra room to write?

Together the classroom teacher and special education teacher review the classroom requirements and the special needs of the students and try to anticipate the accommodations and modifications that best support student learning. Periodically, they discuss the effectiveness of the various approaches; determine which should be continued, are no longer needed, and are not effective; and identify alternative approaches and new areas of need.

In the coteaching model, both teachers will instruct and support. The classroom teacher will often take the lead on content delivery, and the special education teacher might focus on making sure that students have processed the information and organized a response. At other times, the special education teacher will lead the discussion while the classroom teacher observes and evaluates student responses. Both teachers will pose questions that prompt reflection on the mathematical concept. Often, they will help the student recall related ideas that would be helpful in a particular case. Sometimes the special education teacher will give another example for those who need it. If certain prerequisite skills are weak, the special education teacher may devise ways to offer the needed support during the lesson. Special materials may have been devised before the lesson to help the students organize responses. Both teachers will informally assess information during the lesson to determine whether the students are learning. If the students can complete the task, but only with multiple cues, the teachers will recognize the need for further experiences with that particular mathematical concept or skill. The teaching team will use actual performance data and informal assessment information to plan the next lesson. Both teachers must be equal partners in a coteaching environment, which requires mutual respect and adaptability.

Things parents need to know about instruction

Parents can help support their child's education by stressing the importance of mathematics in daily life. Moreover, parents of special-needs students need to begin to learn some of the strategies for modifying instruction and materials that may be easily used at home. Splitting homework into two short periods might work for some students with attention problems. Teachers should share with parents strategies for helping students organize work, as well as effective strategies for memorizing. Teachers should deliver information for students to memorize in small segments, with the gradual addition of unpracticed material. Although initial learning might be in the student's preferred learning style (e.g., oral practice), students should eventually practice information in the format in which they will be required to produce it; if they have written or timed tests in school, then at least some practice should be written or timed.

To effectively assist children, parents need information about the mathematical approaches in the classroom. Many parents of special-needs students want to help but know that teaching a child a method that differs from that of the classroom may only confuse the child. Teachers should give parents examples of appropriate responses so that parents can connect prior learning with the teacher's approach. Examples are needed both when using expanded computation procedures and when asking students to justify problem solutions.

Administrative support

Effective mathematics programs receive administrative support in the form of time, tools, and professional development. Principals hold beliefs about mathematics teaching approaches, and their views and actions can support or derail effective instruction. Teachers need time to prepare for effective mathematics instruction. If students with disabilities are involved, general education and special education teachers need time to collaborate on effective instructional approaches. Administrators must make a strong effort to ensure allocated time for this collaboration and that other scheduled events do not prevent these conversations.

Special education teachers as well as general education teachers need to participate in professional development on the changing approaches in teaching and learning mathematics so that they can stay abreast of the mathematics curriculum and its delivery. Similarly, professional development providers must be held accountable for addressing how these approaches can work effectively for students with special needs as well as for the general student population. If mathematics programs are adopting new tools for mathematics instruction, such as graphing calculators, the principal, as instructional leader, must consider how to furnish these tools and related training to students with learning problems and their teachers. All instructors of mathematics in the school should be part of the dialogue regarding what is important mathematics for all students, what aspects of current instructional approaches present stumbling blocks to students with learning problems, and what systemic changes need to be made. Within this dialogue, everyone needs to hear the voices of teachers who are effective with a wide range of students. Parents should be part of the dialogue so that they can see the relationship between the mathematics curriculum and the future of their children.

Conclusion

To guide our efforts in improving learning of individuals experiencing difficulties in mathematics, let us briefly reexamine and reflect on Bottge's (2001) key–lock model. Recall that the model is based on theories of cognition; emphasizes the NCTM (2000) Equity Principle; and considers learner, contextual, and task variables essential to an adequate description of teaching and learning mathematics. For significant learning to occur, the six teeth of the instruction key (meaningful, explicit, informal, [de]situational, social, and teacher specific) must each fit a pin of the learning lock (engagement, foundations, intuitions, transfer, cultural supports, and student specific). Without attention to the interplay of these variables, students may never have the opportunity to see mathematics as a subject of worth, a subject of connections, a subject that they can truly own rather than simply have borrowed.

This chapter introduces an awareness of the instructional problems and concerns that you may face dealing with students with learning difficulties. The goal is not so much to answer questions as it is to pose them. We intend to offer a base from which you can add options, subtract barriers, multiply chances, and appropriately divide your attention to meet the needs of all students—not simply those who have been labeled for one reason or another. Just as each student has a unique learning approach, we each have a distinctive teaching style. This style is a blend of our personal characteristics, values, experiences, and professional training. By examining aspects of your teaching approach, you can determine congruencies as well as incongruencies between you and the learner.

The chart in figure 3.3 (which appears below and continues across two more pages) discusses and summarizes teaching style, intervention and remediation, accommodation, evaluation, referral, support services, and other aspects of effective mathematics programs for all students. By implementing some of the ideas or adaptations that this chart incorporates, teachers should find themselves in a better position to accommodate the needs of all students—not only those students in the classroom who are experiencing learning difficulties. We hope that this information will serve you well as you strive to meet the day-to-day challenges of instruction.

* * *

Adapting Materials

- White space to separate problems and ideas.

- Increase font size and room to work problems.

- Consistent format for handouts and worksheets.

- Relevant graphics to depict a concept versus graphics with only some connection.

- Materials to help students organize work on a page (examples: division of page into four, six, or eight sections; rotating lined paper ninety degrees to give columns for multidigit operations).

- Materials to block out excess stimuli on a page (examples: use of black construction paper, folding back pages so less is exposed at one time).

- Provide mats as a work surface for working with manipulatives and containers for extra manipulatives.

- Provide a skeletal framework for the steps of an algorithmic computation.

- Recording worksheets separated into stages to accompany a problem-solving activity.

- Rewrite or orally reword directions, explanations, and work problems for students with reading problems. (Peers or aides can also read materials to students having difficulty.)

Fig. 3.3. Instructional strategies for students with special needs

Adapting Instruction

- Presenting concepts for a variety of learning styles (hearing "it," seeing "it," touching "it," manipulating "it," writing "it," or drawing "it").

- Multiple representations of concepts *(example: representing decimals with dollars, dimes, and pennies; place value blocks; and a metric ruler).*
- Use caution to avoid overwhelming with too many representations in the beginning stages of learning a new idea.

- Active engagement with meaningful tasks.

- Model self-questioning and self-monitoring.

- Developing conceptual understanding by beginning with concrete experiences and then moving to pictorial modeling and later abstract representations. *(Explicitly link the concrete and pictorial representations with the symbolic and abstract representations.)*

- Anchored instruction, which presents real-life situations as a context for problem solving *(examples: problems related to a school store setting, video vignettes on discovering measurements embedded in fighting a fire or interpreting medical measurements).*

- Provide challenge without overwhelming. *(Students can be challenged along with other students but may need more support in organizing responses.)*

- Scaffold learning *(examples: rephrasing a problem, modeling a problem solution, talking aloud to show your thinking process, use of questioning to guide students through problem solving).*

- Practice on a skill over many days with short daily reviews.

- Frequent feedback.

- Use of technology to develop concepts, do or check computations, or support problem solving.

- Some explicit instruction during whole-group instruction or with subgroups.

- Incorporate an advance organizer, descriptions and demonstration of concept or skill, guided practice, and a postorganizer.

Adapting Learning Environment

- Extend time needed to learn and practice mathematics. *(If possible, use one-on-one tutoring and instruction from peers, parent and community volunteers, teaching assistants, and others.)*

- Break instruction and assignments into segments to promote on-task behavior.

- Use checklists and checkpoints for longer assignments.

- Mix group and individual activities.

- Rotate cooperative group roles so that all students can fully participate.

- Ensure that all students in a cooperative group have a chance to use language to describe their personal understanding.

Fig. 3.3. Instructional strategies for students with special needs—*Continued*

Adapting Learning Environment

- Ensure that all students in a cooperative group have an opportunity to record problem solution and thinking of the group.
- Incorporate a ten-minute pause and review of goals during cooperative work groups.
- Create a supportive environment where students feel safe taking risks.
- Teach students how to create a supportive environment during group work.
- If using stations, incorporate independent, engaging, and meaningful activities.
- Introduce stations to students as groups.
- Include clear directions and posting of sample work with stations.
- Assign a daily helper to answer questions during station work.
- When creating activities for station work, determine how long students can work independently.
- Use peer tutoring to allow special-needs students to serve as tutors or tutees.

Homework

- Use homework to maintain skill and provide practice.
- Ensure that homework is realistic in length and difficulty.
- Provide clear directions.
- Use meaningful activities closely linked to classroom instruction.
- Check and review completed homework to convey its value.

Teacher Professional Development

- Observations and feedback from colleagues.
- Collaborative model of classroom teacher and special education teacher working together.
- Ongoing professional development for classroom teachers and special education teachers related to mathematics teaching and learning and special-needs students.

Fig. 3.3. Instructional strategies for students with special needs—*Continued*

References

Badian, Nathlie. "Persistent Arithmetic, Reading or Arithmetic and Reading Disability." *Annals of Dyslexia* 49 (December 1999): 43–70.

Babbitt, Bea. "Teaching Geometry Concepts and Skills." In *Designing and Implementing Mathematics Instruction for Students with Diverse Learning Needs*, edited by Pamela Hudson and Susan P. Miller, pp. 489–535. Boston: Pearson, 2006.

Baker, Janice M., and Naomi Zigmond. "Are Regular Education Classes Equipped to Accommodate Students with Learning Disabilities?" *Exceptional Children* 56 (April 1990): 515–26.

Baroody, Arthur J., and Herbert P. Ginsburg. "A Cognitive Approach to Assessing the Mathematical Difficulties of Children Labeled 'Learning Disabled.'" In *Handbook on the Assessment of Learning Disabilities: Theory, Research, and Practice*, edited by H. Lee Swanson, pp. 117–228. Austin, Tex.: Pro-Ed, 1991.

Baxter, Juliet A., John Woodward, and Deborah Olson. "Effects of Reform-Based Mathematics Instruction on Low Achievers in Five Third-Grade Classrooms." *Elementary School Journal* 101 (May 2001): 529–47.

Bottge, Brian A. "Reconceptualizing Math Problem Solving for Low-Achieving Students." *Remedial and Special Education* 22 (March–April 2001): 102–12.

Butler, Frances M., Susan P. Miller, Kevin Crehan, Beatrice Babbitt, and Thomas Pierce. "Fraction Instruction for Students with Mathematics Disabilities: Comparing Two Teaching Sequences." *Learning Disabilities Research and Practice* 18 (May 2003): 99–111.

Carnine, Douglas. "Instructional Design in Mathematics for Students with Learning Disabilities." *Journal of Learning Disabilities* 30 (March–April 1997): 130–41.

Carnine, Douglas, Robert Dixon, and Jerry Silbert. "Effective Strategies for Teaching Mathematics." In *Effective Teaching Strategies That Accommodate Diverse Learners*, edited by Edward Kame'enui and Douglas Carnine, pp. 93–112. Upper Saddle River, N.J.: Merrill, 1998.

Carnine, Douglas, Eric D. Jones, and Robert Dixon. "Mathematics: Educational Tools for Diverse Learners." *School Psychology Review* 23, no. 3 (1994): 406–27.

Cawley, John F., and Rae Reines. "Mathematics as Communication: Using the Interactive Unit." *Teaching Exceptional Children* 28 (Winter 1996): 29–34.

Cognition and Technology Group at Vanderbilt. "Anchored Instruction and Its Relationship to Situated Cognition." *Educational Researcher* 19 (August–September 1990): 2–10.

Council of Chief State School Officers (CCSSO). *Common Core State Standards.* Washington, D.C.: CCSSO, 2010. www.corestandards.org.

Deshler, Donald, Edwin S. Ellis, and B. Keith Lenz. *Teaching Adolescents with Learning Disabilities: Strategies and Methods.* Denver: Love Publishing, 1996.

Doty, Richard G., Susan Mercer, and Marjorie A. Henningsen. "Taking on the Challenge of Mathematics for All." In *Changing the Faces of Mathematics: Perspectives on Latinos*, edited by Luis Ortiz-Franco, Norma G. Hernandez, and Yolanda De La Cruz, pp. 99–112. Reston, Va.: National Council of Teachers of Mathematics, 1999.

Fennema, Elizabeth, Megan L. Franke, Thomas P. Carpenter, and Deborah A. Carey. "Using Children's Mathematical Knowledge in Instruction." *American Educational Research Journal* 30 (Fall 1993): 555–83.

Fox, Mervyn A. "Clumsiness in Children: Developmental Coordination Disorders." *Learning Disabilities: A Multidisciplinary Journal* 9 (Summer 1998): 57–63.

Friend, Marilyn, and Lynne Cook. *Interactions: Collaboration Skills for School Professionals.* 2nd ed. White Plains, N.Y.: Longman, 1996.

Fuchs, Douglas, P. G. Mathes, and Lynn S. Fuchs. *Peabody Peer-Assisted Learning Strategies Reading Methods.* Nashville, Tenn.: Peabody College, 1996.

Fuchs, Lynn S., and Douglas Fuchs. "Principles for the Prevention and Intervention of Mathematics Difficulties." *Learning Disabilities Research and Practice* 16 (May 2001): 85–95.

Fuchs, Lynn S., Douglas Fuchs, Kathy Karns, and Norris Phillips. *Peabody Peer-Assisted Learning Strategies in Math.* Nashville, Tenn.: Peabody College, 1996.

Garrison, Leslie, and Jill Kerper Mora. "Adapting Mathematics for English-Language Learners: The Language–Concept Connection." In *Changing the Faces of Mathematics: Perspectives on Latinos*, edited by Luis Ortiz-Franco, Norma G. Hernandez, and Yolanda De La Cruz, pp. 35–48. Reston, Va.: National Council of Teachers of Mathematics, 1999.

Gersten, Russell, David Chard, Madhavi Javanthi, Scott K. Baker, Paul Morphy, and Jonathan Flojo. "Mathematics Instruction for Students with Learning Disabilities: A Meta-Analysis of Instructional Components." *Review of Educational Research* 79 (September 2009): 1202–42.

Ginsburg, Herbert P. "Mathematics Learning Disabilities: A View from Developmental Psychology." *Journal of Learning Disabilities* 30 (January–February 1997): 20–33.

Greenwood, Charles R., Carmen Arreaga-Mayer, and Judith J. Carta. "Identification and Translation of Effective Teacher-Developed Instructional Procedures for General Practice." *Remedial and Special Education* 15 (May 1994): 140–51.

Greenwood, Charles R., Joseph C. Delquadri, and Judith J. Carta. *Together We Can! Classwide Peer Tutoring to Improve Basic Academic Skills.* Longmont, Colo.: Sopris West, 1997.

Halmhuber, Nancy L., and Scott G. Paris. "Perceptions of Competence and Control and the Use of Coping Strategies by Children with Disabilities." *Learning Disability Quarterly* 16 (Spring 1993): 93–111.

Heddens, James W., William R. Speer, and Daniel J. Brahier. *Today's Mathematics: Concepts, Methods, and Classroom Activities.* 12th ed. Hoboken, N.J.: John Wiley & Sons, 2009.

Hofmeister, Alan M. "Elitism and Reform in School Mathematics." *Remedial and Special Education* 14 (November–December 1993): 8–13.

Hudson, Pamela, and Susan P. Miller. *Designing and Implementing Mathematics Instruction for Students with Diverse Learning Needs.* Boston: Pearson, 2006.

Hutchinson, Nancy L. "Effects of Cognitive Strategy Instruction on Algebra Problem Solving of Adolescents with Learning Disabilities." *Learning Disability Quarterly* 16 (Winter 1993): 34–63.

Keeler, Marsha L., and H. Lee Swanson. "Does Strategy Knowledge Influence Working Memory in Children with Mathematical Disabilities?" *Journal of Learning Disabilities* 34 (September–October 2001): 418–34.

Kilpatrick, Jeremy, Jane Swafford, and Bradford Findell, eds. *Adding It Up: Helping Children Learn Mathematics.* Washington, D.C.: National Academies Press, 2001.

Langone, John. "Managing Inclusive Instructional Settings: Technology, Cooperative Planning, and Team-Based Organization." *Focus on Exceptional Children* 30 (April 1998): 1–15.

Lerner, Janet W. *Learning Disabilities: Theories, Diagnosis, and Teaching Strategies.* 8th ed. Boston: Houghton Mifflin, 2000.

Mastropieri, Margo A., Thomas E. Scruggs, and S. Shiah. "Mathematics Instruction for Learning Disabled Students: A Review of Research." *Learning Disabilities Research and Practice* 6, no. 2 (1991): 89–98.

Miller, Susan P. *Validated Practices for Teaching Students with Diverse Needs and Abilities.* Boston: Allyn and Bacon, 2002.

Miller, Susan, and Cecil D. Mercer. "Educational Aspects of Mathematics Disabilities." *Journal of Learning Disabilities* 30 (January–February 1997): 47–56.

Montague, Marjorie. "Cognitive Strategy Instruction in Mathematics for Students with Learning Disabilities." *Journal of Learning Disabilities* 30 (March–April 1997): 164–77.

Montague, Marjorie, B. Applegate, and K. Marquard. "Cognitive Strategy Instruction and Mathematical Problem-Solving Performance of Students with Learning Disabilities." *Learning Disabilities Research and Practice* 8, no. 4 (1993): 223–32.

National Council of Teachers of Mathematics (NCTM). *Principles and Standards for School Mathematics.* Reston, Va.: NCTM, 2000.

———. *Curriculum Focal Points for Prekindergarten through Grade 8 Mathematics: A Quest for Coherence.* Reston, Va.: NCTM, 2006.

Patton, James R., Mary E. Cronin, Diane S. Bassett, and Annie E. Koppel. "A Life Skills Approach to Mathematics Instruction: Preparing Students with Learning Disabilities for the Real-Life Math Demands of Adulthood." *Journal of Learning Disabilities* 30 (March–April 1997): 178–87.

Paulu, Nancy. "Helping Your Child with Homework." 1995. U.S. Department of Education. www2.ed.gov /parents/academic/help/homework (accessed March 22, 2010).

Resnick, Lauren B., Victoria Bill, Sharon Lesgold, and Mary N. Leer. "Thinking in Arithmetic Class." In *Teaching Advanced Skills to At-Risk Students*, edited by Barbara Means, Carol Chelemer, and Michael S. Knapp, pp. 27–53. San Francisco: Jossey-Bass, 1991.

Richek, Margaret Ann, Joyce H. Jennings, JoAnne Schudt Caldwell, and Janet W. Lerner. *Reading Problems: Assessment and Teaching Strategies.* Boston: Allyn and Bacon, 2002.

Schumaker, Jean B., and Donald Deshler. "Secondary Classes Can Be Inclusive, Too." *Educational Leadership* 52 (December 1994–January 1995): 50–51.

Schumm, Jeanne Shay, Sharon Vaughn, Diane Haager, Judith McDowell, Liz Rothlein, and Linda Saumell. "General Education Teacher Planning: What Can Students with Learning Disabilities Expect?" *Exceptional Children* 61 (February 1995): 335–52.

Scruggs, Thomas, and Margo Mastropieri. "Teacher Perceptions of Mainstreaming/Inclusion, 1958–1995: A Research Synthesis." *Exceptional Children* 63 (Fall 1996): 59–74.

Shalev, Ruth S., Orly Manor, Judith Auerbach, and Varda Gross-Tsur. "Persistence of Developmental Dyscalculia: What Counts? Results from a 3-Year Prospective Follow-Up Study." *Journal of Pediatrics* 133 (September 1998): 358–62.

Vygotsky, Lev S. *Thought and Language.* Cambridge, Mass.: MIT Press, 1962.

Witzel, Bradley, Cecil D. Mercer, and David M. Miller. "Teaching Algebra to Students with Learning Difficulties: An Investigation of an Explicit Instruction Model." *Learning Disabilities Research and Practice* 18 (May 2003): 121–31.

CHAPTER 4

Assessment

Herbert P. Ginsburg
and Amy Olt Dolan

This chapter's goal is to help teachers more effectively assess students with special needs, particularly students who have difficulty learning mathematics. The first sections clarify the nature of assessment, discuss difficulties inherent in doing it, and offer reasons for building it into everyday classroom activities and drawing on test results. The next section describes four major types of assessment: observations, performance assessments, flexible (or clinical) interviews, and tests that teachers can use at different age levels with different types of children. The chapter concludes with discussing how assessment can contribute to the teacher's learning and professional development and how students may benefit from getting involved in assessment themselves.

The Nature, Difficulties, and Goals of Assessment

What is assessment?

Assessment is gaining insight into students' knowledge and motivation. It is discovering what students know and understand, how well they are performing, what they are learning and having trouble learning, and how they interpret and solve problems. Assessment is also learning what students are interested in, what they are feeling, and what motivates them.

Assessment can be formal or informal. Formal assessment usually involves tests of some kind. These may be standardized measures, such as nationally normed standardized achievement tests, state-mandated achievement tests, "cognitive" tests that aim to inform about underlying thought processes, or perhaps classroom tests that a teacher devises to determine whether a lesson or unit has been successful.

But assessment is not limited to testing. Assessment can be informal too. In everyday instruction, teachers give problems to solve and then observe, listen, and talk with students to learn what they know and do not know. Consider the following situations:

- A teacher gives Arthur several two- and three-digit addition problems to solve and determines how many of each type he can get correct.
- A teacher observes that Lisa looks confused when she tries to multiply.
- A teacher sees that Keith always adds both the numerators and denominators of

fractions (thus, $\frac{1}{2} + \frac{1}{2} = \frac{2}{4}$) and therefore consistently gets the same type of in-correct answer.

◆ A teacher listens as Alexandra tells how she got the answer to a problem and then asks questions to clarify her strategy.

All these activities are forms of everyday, in-the-classroom, or formative assessments. Indeed, they are such common and natural ways of interacting with students that teachers do not think to call them "assessment."

The goals of assessment

Several professional organizations concur that assessment can be valuable for teachers in the classroom. "Assessment should support the learning of important mathematics and furnish useful information to both teachers and students" (National Council of Teachers of Mathematics [NCTM] 2000, p. 22). Assessment is even important for preschool and kindergarten children. A recent NCTM position statement advocates that teachers of children at this level should "actively introduce mathematical concepts, methods, and language through a range of appropriate experiences and teaching strategies. These should be monitored by observation and other informal evaluations to ensure that instructional decisions are based on each child's mathematical needs" (2007). Similarly, early childhood experts emphasize the importance of repeated observing, listening, and questioning as effective methods for assessing young children in their everyday context (Copley 1999).

Properly done, both informal or formative and formal assessment can give teachers reliable and accurate information that can yield several benefits:

◆ Most important, assessment can contribute to more effective instruction for all children, including those with learning difficulties. Good teaching elaborates on, expands, and systematizes students' informal knowledge and everyday interests. It is responsive to students' efforts at learning, to their understandings and confusions. Good teaching, in short, depends on understanding students' thinking and learning—that is, on sensitive assessment (Ginsburg 2009).

◆ Assessment can help identify special difficulties that students sometimes face in learning. One child may have trouble dealing with verbal material and another with remembering number facts. Assessment can inform teachers about reasons why students, particularly students with special needs, have trouble learning.

◆ Assessment can produce useful information to communicate with parents or other professionals who may then work with and help the student.

◆ Assessment can help the student understand and overcome his or her own learning problems. This often happens when the student gains insight into unsuspected strengths as well as difficulties.

Role of Assessment in Diagnosing a Learning Problem

Assessments, both formal and informal, play an important role in identifying that a learning problem exists. The classroom teacher, school psychologist, speech and language pathologist, occupational therapist, educational diagnostician, and parents work together to gather information to determine what areas of learning the problem affects and whether the student is eligible for special education services as defined by the Individuals with Disabilities Education Act (IDEA).

If the child is identified with a specific learning disability, the assessment data aid in developing an individualized education program (IEP). The IEP is a written agreement between the parents and the school that summarizes the child's learning needs and how the school will meet those needs. The IEP includes the following (Hallahan and Kauffman 1994):

- Current level of academic performance
- Annual goals for the student
- Instructional objectives related to the annual goals
- Special educational services to be offered, including accommodations or modifications
- Expected duration of services
- Plans for measuring student progress and assessing, at least annually, whether the goals and objectives are being met
- Transition planning for older students

Barriers to mathematics learning

Students with a learning disability have trouble receiving, processing, storing, or responding to information. They are of average or above-average intelligence but show underachievement in actual performance. Some characteristics of learning disabilities that can contribute to underachievement in mathematics are as follows (Allsopp et al. 2003):

- Memory problems—difficulty learning mathematics facts or remembering the steps for an algorithm or mathematics procedure
- Auditory processing problem—difficulty processing oral explanations of mathematics content
- Visual processing problems—difficulty processing and differentiating mathematics content visually (e.g., confusing numbers such as 25 and 52, copying work inaccurately, losing place while working problems, working problems in the wrong direction, reversing negative and positive numbers on a graph, lining up numbers in the wrong place)
- Motor processing problems—may include difficulty in writing numbers or writing legibly in small spaces, leading to mathematical errors

◆ Abstract reasoning problems—difficulty with problem solving and learning new concepts

◆ Organizational problems—difficulty organizing time, materials, or information

◆ Attention problems—difficulty focusing on important information or attending to it meaningfully

State Assessment and Special Education

In 2001, Congress passed the No Child Left Behind (NCLB) Act. NCLB requires annual testing in specific academic areas and grades. For mathematics, students will be tested each year during grades 3–8 and once during grades 10–12.

Students who require special education services are not exempt from this testing. However, these students are allowed the appropriate accommodations necessary to participate in the tests. These accommodations do not change what the test measures but allow students to show what they know by removing the learning barrier that the disability causes. Each child's IEP specifies the appropriate accommodations. Typical accommodations can be categorized into presentation, response, timing and scheduling, and setting accommodations (fig. 4.1).

Some difficulties with assessment

Despite its laudable goals, many teachers are leery of formal assessments such as those that states and NCLB require. First, they tend to think of such assessments as formal, pressured testing that does little good. Teachers feel that whatever its benefits, high-stakes assessment—mandated achievement tests of some type—forces them to teach according to the test, does not give them information useful for understanding students or improving teaching, makes students anxious, and takes time from the major focus of the classroom: teaching mathematics.

Despite potential benefits (e.g., evaluating program effectiveness or ensuring that children get adequate education), high-stakes testing can also result in negative outcomes for students and teachers alike. It can be stressful and produce great discomfort in students, sometimes preventing them from performing as well as they might. Students with disabilities often perform more poorly than their nondisabled peers on such tests. These tests often fail to offer teachers insight into student learning; fail to improve classroom instruction; and, indeed, distort mathematics education. For these reasons, and because no single measure alone can accurately reflect the knowledge and skills of a student, classroom-based daily assessment is important. Teachers can learn to conduct and benefit from informal or formative assessment in the classroom: observations, performance assessments, interviews, and tests. Teachers can also learn from certain tests that school psychologists and other assessment specialists give. We consider next some major types of informal and formal assessment.

Presentation accommodations

Who can benefit	Questions to ask	Examples
Students with print disabilities, defined as difficulty or inability to visually read standard print because of a physical, sensory, or cognitive disability	• Can the student read and understand directions? • Does the student need directions repeated frequently? • Has the student been identified as having a reading disability?	• Large print • Magnification devices • Human reader • Audio tapes • Screen reader • Talking materials (calculators, clocks, timers)

Response accommodations

Who can benefit	Questions to ask	Examples
Students with physical, sensory, or learning disabilities (including difficulties with memory, sequencing, directionality, alignment, and organization)	• Can the student use a pencil or other writing instrument? • Does the student have a disability that affects spelling ability? • Does the student have trouble with tracking from one page to another and maintaining his or her place?	• Scribe • Notetakers • Tape recorder • Respond on test booklet • Spelling and grammar devices • Graphic organizers

Timing and scheduling accommodations

Who can benefit	Questions to ask	Examples
Students who need more time, cannot concentrate for extended periods, have health-related disabilities, fatigue easily, or have special diet or medication needs	• Can student work continuously during the entire time allocated for test administration? • Does student tire easily because of health impairments? • Does student need shorter working periods and frequent breaks?	• Extended time • Frequent breaks • Multiple testing sessions

Setting accommodations

Who can benefit	Questions to ask	Examples
Students who are easily distracted in large-group settings, concentrate best in small groups	• Do others easily distract the student? • Does student have trouble staying on task? • Does student exhibit behaviors that would disrupt other students?	• Change of room or location of room • Earphone or headphones • Study carrels

Fig. 4.1. Determining appropriate assessment accommodations for students with disabilities (adapted from National Center for Learning Disabilities 2005)

Classroom-Based Formative Assessments

No one type of assessment can give all the information teachers need about a student. Teachers should use several methods to get to know what a student is thinking and learning. These methods include observation; performance assessment; flexible (or clinical) interview; and tests, both informal and formal. All can offer useful information.

Observation

Assessment should begin with observing ordinary student behavior: what they do, what they say, what kind of answers they give in class or on examinations, and what kind of homework they produce. The first thing to observe is overt performance and correctness of response. The need to keep track of overt achievement is obvious, and doing so is fairly easy. The teacher cannot wait for the result of some achievement test to learn whether students are succeeding or failing. The teacher needs to observe in the classroom how well students calculate, solve various problems, and generally deal with the current material.

But to furnish deeper insight, observation needs to focus on more than correctness of response. Indeed, the goal should be to look beneath overt behavior to uncover underlying processes. Observation should attempt to detect patterns underlying student errors, strategies determining responses, and other mental and motivational processes that can affect student learning. Here is an example involving students' use of the blackboard.

Blackboard mathematics

Kay Kobe, a third-grade teacher, began by giving her students a problem that they had not seen before: 23 × 4. The students had previously worked on single-digit multiplication facts, but no more than that. Next, she asked the students to solve the unfamiliar problem in any way desired and to write down as much of the method of solution as possible. As students worked, she circulated around the room, observing what they wrote.

Although this particular problem was new, the teacher's method was not. She often encouraged students to work problems on their own and to write and talk about their methods of solution.

When the students were finished with their solution to 23 × 4, the teacher selected several children to write their method on the blackboard for the entire class. Her choices were not random: she chose the children to illustrate different methods of solution. Figure 4.2 shows four methods that the children used that day. (For the full example, see Ginsburg and Seo 1999.)

Method 1 was the most developmentally primitive and incomplete as well. The child began by creating seven vertical groups of three circles and then a vertical group of two circles. Then he wrote the equals sign followed by the written number 23. He repeated this entire procedure only once; stopped there, apparently overwhelmed by the task of counting up all the units; and did not achieve a solution.

In methods 2 and 3 the children transformed 4 × 23 into 23 + 23 + 23 + 23. One child then added pairs of 23s to obtain two 46s, which the child then added to obtain the sum. The other child added three 23s to obtain 69, which the child then added to 23.

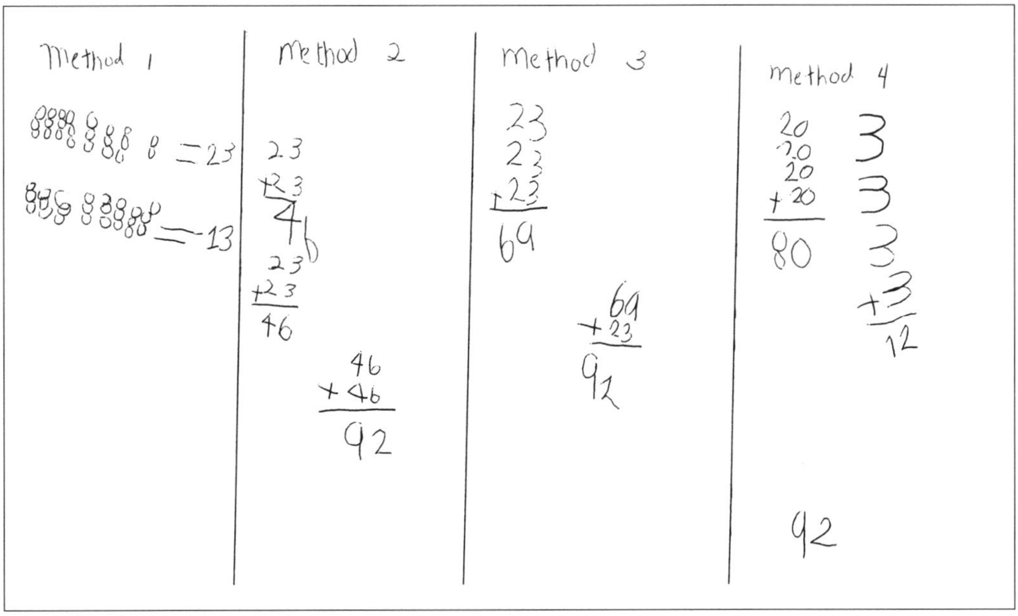

Fig. 4.2. Four methods for solving 23 × 4

In method 4, the child explicitly converted 23 into the combination of 20 and 3 (23 is "really" 20 and 3), added the tens and units to obtain their sums, and then added the sums of the tens and the sums of the units to obtain the overall sum. Thus the child essentially did the following in sequence:

1. $23 + 23 + 23 + 23 = (20 + 3) + (20 + 3) + (20 + 3) + (20 + 3) =$
2. $(20 + 20 + 20 + 20) + (3 + 3 + 3 + 3) =$
3. $80 + 12 = 92$

The example makes several important points. First, the basic task was easy for the teacher to create. Children used chalk to write methods of solution to an arithmetic problem on the blackboard. Most any classroom could replicate this situation.

Second, although the task was easy for the teacher to create in the classroom, the problem 23 × 4 was not trivial to the students. She asked the children to use prior knowledge to solve a problem they had never seen before and then carefully chose certain solutions for extended observation. Had she chosen a trivial problem and not encouraged the children to write methods of solution, their overt behavior would have revealed little of interest. Much, perhaps even most, overt behavior in the classroom is not interesting; teachers must give students carefully selected examples.

Third, the children's correct answer was the least interesting aspect of their behavior. Clearly, the methods of solution that led to the correct answers gave the teacher enormous insight into the children's knowledge. For example, the teacher saw that method 1 was rather primitive in that it involved counting by ones, an unwieldy and inefficient approach. Methods 2 and 3, however, share the basic and common idea that one can think of multiplication as successive addition. Further, in both cases, the children knew that they could combine a sequence of numbers in convenient ways for efficient addition. Method 4 was even more interesting, introducing the idea of the base-ten composition of numbers, and is an application of the distributive property, which is similar to the standard algorithm.

Fourth, the observations laid the basis for profitable lines of instruction. The teacher could help the child using method 1 to realize that counting by ones, although it leads to a correct answer, is not an efficient way to multiply. She helped that child to appreciate the power of grouping. She also saw from methods 2 and 3 that many children were ready to use ideas of grouping to understand multiplication. She saw further that method 4 supplies a nice bridge between informal grouping methods, use of the distributive property, and the standard algorithm. If the child can understand that adding 23 four times is the same as adding four 20s and four 3s, then introducing the algorithm, in which one first multiplies 3 by 4 and then 20 by 4 to get the product, will not be hard.

Seatwork and classroom tests

Blackboard mathematics is not the only everyday classroom practice that affords making useful observations about children's mathematical thinking. Almost every teacher above the preschool and perhaps kindergarten levels gives students seatwork and tests. One aim, of course, is to determine how well students have mastered the material (being able to solve a reasonable number of relevant problems). But observation of seatwork and tests can also yield insight into underlying processes. The teacher can observe how the students solve problems and can examine their written work. The teacher needs to focus not only on whether the child gets the right answer but also on the child's approach or strategy. What does the student write down first? Did the student calculate the individual numbers accurately or write them down in the wrong place? What numbers did the student cross out? The student's actions and written work can give valuable clues about thinking. But of course the clues do not scream out the interpretation that the teacher seeks, just as the mathematical problem does not proclaim its answer to the student. The teacher needs to consider what the clues mean, just as the student needs to think about how to solve the mathematical problem.

Performance assessment

Observation is hard to do. It demands that the teacher focus on and immediately interpret ongoing events, some of which one cannot anticipate. Performance assessment has observation at its core but is more organized and simpler to use. Typical performance assessment involves presenting students with at least one "authentic" problem that the curriculum has emphasized and then, over time, documenting students' classroom performance—not test performance—through methods such as checklists and portfolios of student work. After repeated performance assessment involving multiple observations of a student, enough evidence is available to assess the student's overall strengths and weaknesses (Bowman, Donovan, and Burns 2001, p. 249). In contrast to standardized testing, performance assessment does not directly compare performance across students but attempts instead to inform concerning an individual student's performance in relation to curriculum goals. The issue is not whether Kyoung-Hye performs better than most students at her grade level but rather what exactly she has learned about what the teacher is trying to teach.

Over the past decades, researchers have developed and refined several approaches to performance assessment. For example, at the preschool and kindergarten levels, the *Big Math for Little Kids* curriculum (Ginsburg, Greenes, and Balfanz 2003) offers repeated opportunities for performance assessment of number, operations, measurement, shape, spatial relations, logic, and pattern. The *Work Sampling System* (Meisels et al. 1994) is appropriate for children from preschool through grade 5. Checklists, developmental guidelines, portfolios, and summary reports are the crux of this system. Teachers use these tools to examine how individual students interact in the classroom with materials, adults, and peers, and from these observations teachers can learn something about what students achieve and how to teach to them.

At the middle and high school levels, Katims and his colleagues (1994) developed a fascinating collection of mathematical activities suitable for performance assessment. The Summer Jobs problem tells small groups of middle school students the following: last summer, Maya had a concession business at an amusement park. Nine vendors sold popcorn and drinks to whomever they could in the park; they worked different numbers of hours on different days. This summer, Maya can hire only six vendors, three full time and three half time, and she wants to rehire the most productive vendors from last year. Maya reviewed last summer's monthly sales records for each vendor. But how can she figure out which vendors were most productive if they worked different numbers of hours in different places? Small teams of students work as a group to solve such problems. They discuss their methods, keep written records, and in the end must write a letter to Maya giving a solution and explaining how they evaluated the vendors.

Given such problems, the students not only engage in rich learning, but their overt behavior—their discussions, their written materials, and finally their written solution and justification for it—also reveals their thinking about the mathematical issues underlying the problem (here, statistical considerations).

Flexible interview

Observation and performance assessment may produce valuable insights. But often the observations are not conclusive. The teacher wonders whether the student who solves the problem "really" understands the basic concepts or whether the child who uses one strategy could just as easily use another. The teacher is not certain whether the student who fails in solving a problem could in fact solve it with a little help. The letter to Maya is not entirely clear with respect to students' understanding of a particular statistical idea, and the teacher wonders what a certain statement in the letter really meant. Observations alone may not furnish enough information to answer these questions.

One solution is to engage the child in what Piaget (1976) called a "clinical interview," or what we refer to as a "flexible interview." The essence such an interview is its flexible, responsive, and open-ended nature. (For a more complete description of the interview method, see Ginsburg [1997]). In the flexible interview, the interviewer asks the student to reflect on and articulate his or her thinking processes. The interviewer establishes rapport, prepares a series of appropriate tasks, asks questions to uncover how the student thinks, and listens to and observes the student's responses. The observant and thoughtful interviewer constantly adapts interviewing strategy to the direction that the student suggests.

Nonspecific questions such as "How did you do it?" or "What did you say to yourself?" or "How would you explain it to a friend?" encourage rich verbalization but give no suggestion of how the student should respond. The interviewer should encourage students to work with chips or tiles, pencils or marker pens and paper, and small toys. These manipulative objects motivate the student, and because they tend to externalize thinking they allow the interviewer–observer more insight into thinking processes.

Responding to the student's responses and behavior, the interviewer makes and tests hypotheses about the student's thinking. The interviewer varies and modifies tasks, which become more specific to focus on particular aspects of thinking and more difficult to test the limits of the student's understanding.

The interviewer is not satisfied with a "correct" answer but instead tests the strength and consistency of the student's beliefs with repetition and countersuggestion. An "incorrect" answer may reveal valuable information about thinking processes. Whether correct or incorrect, the most desirable answer is one that produces the most information about the student's thinking.

Flexible interviewing is nonstandardized in many ways. For some students, establishing rapport requires warm encouragement and calming of apprehension. For others, a tone of high expectation produces better results. Whereas an interviewer can allow one student to follow a train of thought at length, another student will need closer control, with more frequent questions. The interviewer must phrase questions so that the student understands them. A skilled interviewer is aware of the student's personal vocabulary and can rephrase questions in the student's words.

Maintaining interest is necessary to measure the student's best, rather than ordinary, performance. The bored, distracted, tired, or uncomfortable student will not reveal much useful information. Rewarding dialogue happens when the student is interested in the issues, enjoys the attention of the interviewer, and finds the task challenging but not overwhelming. The interviewer may learn not only about the student's thinking but also about important attitudes and beliefs concerning the student's own ability, as well as the goals, methods, and nature of mathematics.

Here is an example of an elementary school–level interview. The interviewer asked Sara, a seven-year-old second grader, "How much is three plus four?" She used her fingers, trying to hide her actions from the interviewer, and obtained the answer "seven." Then the interviewer asked her to explain how she got the answer.

Sara: What you do is this trick: you have three and you know that it's in your head and then on your fingers you count four, five, six, seven. You only have to use four fingers and not all of them, like three of them and then four. So you can see on your fingers that it is four.

From this explanation, the interviewer learned a great deal concerning Sara's understanding of addition. She did not produce the answer by rote memory; instead, she used the strategy of counting on from the first number. In this strategy, she can represent the first number mentally; the second physically, by the fingers; and the sum by the counting words. She knew that she can use counting to solve number combination questions, but she did not yet realize that counting from the larger number would be easier. Sara's answer reveals a good deal of complexity behind her response to one of the simplest aspects of arithmetic, a small-number combination problem.

Later, Sara undertook a simple verbal subtraction problem, "How much is two take away one?" and answered, "Two take away one equals one."

Interviewer: What does "equals" mean?

Sara: Equals means that you are giving your problem an answer. If you have a math problem, say, three take away two, and that's very easy because the answer is one, but if you just put one there, the teacher would think your answer is 21. [*She wrote "3 – 2 1."*] So you need to put some sign to say that these numbers equals the answer.

Interviewer: Is there another word you could put there?

Sara: No, equals is a sign in math, and you can't put a word in there or the teacher would think you didn't understand that you are doing math.

This exchange reveals a good deal about Sara's understanding of the equals sign. First, she does not seem to think of "=" as indicating equivalence; instead, she has an "action interpretation" in which "=" indicates that the student is supposed to get an answer to the stated problem. Moreover, an important function of the equals sign is to separate the written numerals so that one can distinguish between the second addend and the sum. Also, the sign is part of a distinctive language, the language of mathematics, which must be spoken in class and which is not readily translatable into English.

This, then, is what we mean by children's thinking: not just the right or wrong answers to the various problems, but the reasoning, beliefs, and theories of the sort Sara displayed. The goal of assessment should be to discover and understand thinking in order to foster students' learning and ameliorate their difficulties.

Flexible interviewing has a special role for students experiencing difficulties with school mathematics. Here is an example of an interview with a seventh-grade girl who was labeled with a specific learning disability, ADD (attention-deficit disorder). Lisa was asked to solve for x in the ratio problem $x/5 = 2/10$.

Lisa: OK, x divided by 5 and 2 divided by 10. I don't know x, so I'll do this part [*she enters 2/10 into her calculator and writes ".2" on the line below the original problem*].

Interviewer: What is this problem asking?

Lisa: To divide: this over this, and this over this [*pointing to the fractions on both sides of the equals sign*].

Interviewer: What will that tell you?

Lisa: The answer.

Interviewer: Well, you did that, divided on this side, and got .2. Is that your answer?

Lisa: Um, well, sure, that's what I would put.

Interviewer: If I told you this was a ratio, how could you read it to help find a way to solve it?

Lisa: What's a ratio?

The interview revealed that, like Aaron's magical math (chapter 2), Lisa's approach was entirely procedural and did not appear to involve basic understanding of the nature of the problem. In this respect, although classified as ADD, Lisa was no doubt similar to many other students at this level. Indeed, from an instructional point of view, the valuable information for the teacher is not likely to be Lisa's diagnosis (which may or may not be correct) but rather the specific nature of her difficulty with the ratio problem, for this information can lead to a plan for helping her. The interview suggested that in isolation, drill on the correct steps for solving the problem would be unlikely to succeed because she seemed to be missing core areas of understanding. Instead, a fruitful instructional plan should involve helping her understand the concept of ratio as the relationship of one quantity to another and then to appreciate the relation between the concept and the procedures for calculation. In this way, the methods used would make some sense and Lisa would have a way of thinking about how to handle new ratio problems.

Flexible interviews can be powerful tools offering rich insight into students' thinking. Teachers often agree with this proposition, yet they wonder whether interviewing in the classroom is practical. It seems to be. Given some assistance, teachers manage to develop distinctive ways of integrating flexible interviewing into their classrooms and styles of teaching (Ginsburg, Jacobs, and Lopez 1998). For example, one teacher decided to conduct interviews with small groups of children while the rest of the class did seatwork. Another teacher built interviews into group lessons, always asking the children how they solved problems. Another teacher took children aside for individual interviews during recess or lunchtime. Still another teacher trained her students to interview each other and discuss the results with the class.

Interviews need not be difficult to implement in the classroom during instruction if the teacher commits to focus instruction not on the mathematics itself but on how children think about the mathematics. Here are two suggestions about how to begin:

1. Ask several members of the class to do some blackboard mathematics, as described earlier, in which they write on the board as much of their solutions as possible. Then in front of the whole class, interview the students about how they solved the problems. Help them to expand on what they have written on the board and why they solved the problems as they did. Invite other students in the class to ask questions as well. In this manner, blackboard mathematics will become an occasion for all students to think about ideas and methods for solving problems. That is not only assessment but also real mathematics education.

2. Sit down with individual children for a few minutes to review their understanding of the textbook. You can do this as the rest of the class is engaged in seatwork. Ask the child to explain to you the ideas that the textbook is designed to teach. Ask what the book means when it defines a concept; ask for the student's interpretation of the textbook explanation. Ask what certain problems are asking the student to do. Teachers often assume that textbooks are clear, that they convey what they are intended to convey, and that students interpret textbook explanations the way the teacher does. Individual interviews may reveal that sometimes none of these assumptions is true and may offer insight into students' idiosyncratic understanding of textbook material.

Tests

The traditional approach is to assess students by administering tests, both informal (e.g., classroom tests that teachers devise) and formal (e.g., annually administered statewide achievement tests and other standardized tests from the school psychologist or other professional). The essence of a test is that all children receive the same set of questions or test items in basically the same way. Responses are not difficult to score, and computing and comparing test results is easy. Thus, in a classroom test, you can easily determine that most children got almost all the problems correct (they get an A), whereas a few children got about half correct (a B or C), and so on. Such information may tell you something useful about overall level of performance. Similarly, a standardized achievement test, which is based on norms, may reveal that a particular child scores in the lowest 5 percent (fifth percentile) of all children in the district, or even the country.

Some criticize such tests because they fail to treat students as individuals. Suppose that a particular child scored badly because he did not understand the particular wording of the question or because he was not paying attention. Although tests are not perfect and do indeed suffer from such deficiencies, sound justification for using them exists, particularly when other methods, such as observation, performance assessment, and flexible interview, supplement them.

Indeed, one can justify testing on ethical grounds, particularly with respect to a specific kind of fairness. As far back as 1845, Horace Mann offered several reasons for introducing standardized testing to the schools: standardized tests are impartial, are just to the pupils, prevent officious interference by the teacher, take away all possibility of favoritism, make the information obtained available to all, and enable all to appraise the ease or difficulty of the questions (Wainer 1992, p. 15).

In some way, all these reasons involve fairness or impartiality. Tests treat all children the same, ignoring their individuality, in order to be fair. Giving all children the same questions prevents the teacher from favoring some children over others (perhaps by giving some children easier questions than others) or from interfering with the process of testing. Another justification is that the tests make the process public, so that an outside observer could see whether the questions were too hard or easy. In this way, testing can be fair and just. Imagine the complaints were the teacher to give a different test to every child.

Standardized tests for mathematics assessment are generally of two types: achievement and cognitive. Achievement tests basically assess students' overall performance in particular areas of mathematics. Formal achievement tests generally involve norms—data describing the performance of many students, even a nationally representative sample—to make an individual student's performance comparable with that of others. After teaching a certain amount of material, the teacher often devises informal achievement tests to determine students' performance. Both types of tests can generate valuable information. After all, knowing who is and is not doing well, and what material is and is not being mastered, is important. But neither informal nor formal achievement tests are particularly useful for improving teaching, aside perhaps from indicating which topics need further attention. Knowing a student's standard score, percentile rank, or grade-level equivalency on a particular measure is not enough. Such information yields only comparative benchmarks, not the specific information necessary to help the child learn a particular concept(s).

More informative for the teacher are cognitive tests, which aim to inform concerning underlying mathematical knowledge, concepts, and strategies. Ordinarily, teachers do not themselves give these tests. Instead, school psychologists or other assessment specialists give them to individual children in an effort to understand the roots of their learning difficulties. Yet teachers should know what tests to ask for and understand the tests that are administered because they can furnish useful information for improving instruction and developing IEPs. Hence we describe here some major cognitive tests at different age levels.

Early childhood

Few cognitive tests are available for young children, particularly of preschool age. Among them are the K-SEALS (Kaufman and Kaufman 1993), which is appropriate for children aged from three years and zero months to six years and eleven months and offers short number naming, number recognition, and number concept items. The WIAT-II (Wechsler 2001) includes two subtests, mathematics calculation and mathematics reasoning, normed for ages four years through adulthood. The Woodcock–Johnson test (Woodcock, McGrew, and Mather 2001) spans the range from two years to adulthood and contains subtests dealing with calculation, math fluency, and applied problems.

The most extensive cognitive test for children from ages three to eight years is the Test of Early Mathematics Ability, Third Edition (TEMA-3; Ginsburg and Baroody 2003). The TEMA-3 yields information concerning students' functioning in key areas of mathematical knowledge, both formal and informal. The test items were derived from the considerable research literature on mathematical thinking over the past thirty years (Ginsburg, Klein, and Starkey 1998) and deal with such topics as concepts of relative magnitude, the mental number line, cardinality, counting and enumeration, part–whole relations, concrete and mental addition and subtraction, reading and writing numerals, number facts, written calculation, and base-ten concepts.

The TEMA-3 is individually administered and includes both concrete and written problems. The evaluator can use the results to rank the child overall in relation to peers (on the basis of national norms); can obtain separate scores describing the child's proficiency in informal versus formal mathematics; and can examine the child's performance in such areas as informal and formal concepts and procedures. The test lets the examiner identify both strengths and weaknesses. In fact, many testers find that although weak on certain formal skills, such as knowledge of number facts or proficiency with written addition, a child may nevertheless show impressive informal knowledge of basic ideas of cardinality and mental addition. Aware of these strengths, the teacher can then design instruction to build on them.

In brief, the TEMA draws on current research to examine the child's relative performance, under standardized testing conditions, in vital areas of mathematical thinking. Its most important function is to encourage evaluators, as well as the teachers and others who receive the evaluations, to think differently about children's mathematical thinking. The nature and content of the test help the evaluators and others to consider that mathematical thinking has both informal and formal components; that even though weak in formal knowledge or considered "learning disabled," the child may nevertheless possess considerable informal knowledge; that the child may use strategies and concepts to solve problems; and that in general mathematical thinking involves much more than memorizing the number facts or mastering calculational procedures.

Although the TEMA-3 produces useful information, testers and teachers often want to learn even more about a particular child's thinking. Consequently, Ginsburg (2003) developed an organized system of probes to be used in parallel with the TEMA-3. The general idea was that after having given the TEMA-3 in the standard fashion, many examiners would find probing further into the thought processes that produced the observed performance useful, particularly for errors. Most evaluators, however, have not had training or experience in assessing children's thinking; they are not skilled in flexible interviewing. Consequently, Ginsburg attempted to give examiners a structured and comfortable procedure for probing the strategies and concepts underlying children's responses to the TEMA. We can consider such probes a gentle first step toward more extensive flexible interviewing.

The probes for each item of the TEMA-3 involve three main features. The probes first attempt to establish whether the child has understood the basic question. Often children produce an incorrect response because they have misinterpreted a minor feature of the question. The probes attempt to distinguish this situation from that in which children miss the question because they fail to comprehend the relevant concept. Next, the probes attempt to determine the child's strategies and processes to solve the problem. For example, for mental addition, the probes attempt to determine whether the child used such procedures as finger counting, mental counting on, or memorized number facts. Third, the probes attempt to establish learning potential. The issue is whether the child can learn the relevant material with few hints or whether more substantial teaching is required. In the first case, the child is clearly close to "understanding"; in the second case, the child is not. Further, after describing how to establish the child's level of understanding, the manual then recommends educational activities relevant for the material that each TEMA-3 item tested.

Teachers can benefit from probes in at least two ways. First, teachers should request that school psychologists or other assessment specialists use the probe method to follow up on TEMA-3 testing and report the results. Second, an even more valuable procedure would be for teachers to administer the probes themselves. No reason exists why teachers could not use the probes as well as school psychologists or other assessors. Doing so would give teachers useful insights into their students' thinking.

Older students

Several developed tests assess older students' mathematical abilities. As mentioned, the WIAT-II (Wechsler 2001) measures mathematics calculation and mathematics reasoning in students from ages four years through adulthood. The Woodcock–Johnson test (Woodcock, McGrew, and Mather 2001) spans the range from two years to adulthood and contains subtests dealing with calculation, math fluency, and applied problems.

Another test for this age level is the KeyMath-3 Diagnostic Assessment (Connolly 2008), which covers pre-K–grade 9 and focuses on three areas of mathematical knowledge: basic concepts, operations, and applications. The Basic Concepts subtests include numeration, algebra, geometry, measurement, and data analysis and probability. The Operations subtests deal with mental computation and estimation, addition and subtraction, and multiplication and division, all covering models, basic facts, algorithms, and calculations with whole and rational numbers. Finally, the Applications subtests deal with foundations for problem solving and applied problem solving. The KeyMath-3 Diagnostic Assessment is a comprehensive instrument.

For ages eight years and older, one can administer the Test of Mathematical Abilities (TOMA-2; Brown, Cronin, and McEntire 1994). This tool generates information regarding computation skills, math vocabulary, story problems, and general information about math as used in everyday life, as well as a measure of a child's attitude toward math.

The Diagnostic Achievement Battery (DAB-3; Newcomer 2001) assesses various achievement skills, as Public Law 94-142 delineates, including mathematics and written and spoken language. The mathematics subtests measure both reasoning and calculation. The DAB-3 is supplemented by a system of probes (Newcomer and Ginsburg 2001; similar to that accompanying the TEMA-3) that shed light on comprehension of the problem, strategies, and learning potential.

Students and Teachers

Students can benefit in several ways from assessment. When a teacher in the clinical interview asks students to describe their thinking, students learn several important lessons. One is that the teacher considers their thinking to be of major interest. In effect, the teacher is saying, "I value your thinking, your way of dealing with problems. I am interested in your thinking; it is the center of my attention. Your thinking is so important that I must understand it so that I can teach you how to think better." As a corollary, a second lesson is that thinking is a vital part of the curriculum. Students come to see that learning mathematics is learning to think *about* mathematics, not just memorizing facts or procedures. A third lesson is that students need to reflect on and communicate their own thinking. They need to understand how they solve problems and need to learn to describe their thinking in ways that others can understand. So participating in assessment can help students to think differently about their learning of mathematics.

Teachers learn important lessons too. As we have seen, assessment is not primarily a matter of assigning scores or grades to students. Instead, assessment is gaining insight into students' thinking and learning. It is coming to understand students' concepts, ideas, strategies, misconceptions, and ways of thinking about mathematics. Assessing students' thinking requires that the teacher think critically, using imagination to interpret observations, performance assessments, interviews, and tests. Assessment is using the evidence that these assessment procedures generate to create a practical theory of the individual student's behavior, a theory that can guide instruction. In this way, assessment can help teachers learn to think more deeply about student learning.

Further, teachers engaged in meaningful assessment may also reconsider the goals of mathematics education. If the teacher focuses on thinking and students' making sense of the mathematics they are learning, then the mathematics curriculum is not mathematics as it is sometimes conceived, namely, a collection of topics. Instead, the curriculum is *thinking* about the mathematics being learned. In this approach, mathematics teaching involves helping students to think mathematically and to think *about* their mathematical thinking. This thinking-and-sensemaking approach requires a constant focus on (and assessment of) thinking, not on mathematics as a finished, static product. Thinking assessment requires the thinking curriculum, and the thinking curriculum requires thinking assessment.

References

Allsopp, David, Louann Lovin, Gerald Green, and Emma Savage-Davis. "Why Students with Special Needs Have Difficulty Learning Mathematics and What Teachers Can Do To Help." *Mathematics Teaching in the Middle School* 8 (February 2003): 308–14.

Bowman, Barbara T., M. Suzanne Donovan, and M. Susan Burns, eds. *Eager to Learn: Educating Our Preschoolers*. Washington, D.C.: National Academies Press, 2001.

Brown, Virginia L., Mary Cronin, and Elizabeth McEntire. *Test of Mathematical Abilities*. 2nd ed. Austin, Tex.: Pro-Ed, 1994.

Connolly, Austin J. KeyMath-3 Diagnostic Assessment. Minneapolis, Minn.: Pearson, 2008.

Copley, Juanita V., ed. *Mathematics in the Early Years*. Reston, Va.: National Council of Teachers of Mathematics, 1999.

Ginsburg, Herbert P. *Entering the Child's Mind: The Clinical Interview in Psychological Research and Practice*. New York: Cambridge University Press, 1997.

———. *Assessment Probes and Instructional Activities for the Test of Early Mathematics Ability-3*. Austin, Tex.: Pro-Ed, 2003.

———. "The Challenge of Formative Assessment in Mathematics Education: Children's Minds, Teachers' Minds." *Human Development* 52 (April 2009): 109–28.

Ginsburg, Herbert P., and Arthur J. Baroody. *The Test of Early Mathematics Ability*. 3rd ed. Austin, Tex.: Pro-Ed, 2003.

Ginsburg, Herbert P., Carole Greenes, and Robert Balfanz. *Big Math for Little Kids*. Parsippany, N.J.: Dale Seymour Publications, 2003.

Ginsburg, Herbert P., Susan F. Jacobs, and Luz Stella Lopez. *The Teacher's Guide to Flexible Interviewing in the Classroom: Learning What Children Know about Math*. Boston: Allyn and Bacon, 1998.

Ginsburg, Herbert P., Alice Klein, and Prentice Starkey. "The Development of Children's Mathematical Knowledge: Connecting Research with Practice." In *Handbook of Child Psychology: Vol. 4. Child Psychology in Practice*. 5th ed. Edited by Irving E. Sigel and K. Ann Renninger, pp. 401–76. New York: Wiley and Sons, 1998.

Ginsburg, Herbert P., and Kyoung-Hye Seo. "The Mathematics in Children's Thinking." *Mathematical Thinking and Learning* 1 (June 1999): 113–29.

Hallahan, Daniel P., and James M. Kauffman. *Exceptional Children: Introduction to Special Education*. 6th ed. Boston: Allyn and Bacon, 1994.

Katims, N., R. Lesh, B. Hole, and M. Hoover. *PACKETS: A Program for Integrating Learning and Program Assessment for Mathematics*. Lexington, Mass.: Heath, 1994.

Kaufman, Alan S., and Nadeen L. Kaufman. *K-SEALS: Kaufman Survey of Early Academic and Language Skills*. Circle Pines, Minn.: American Guidance Service, 1993.

Meisels, Samuel J., Judy R. Jablon, Dorothea B. Marsden, Margo L. Dichtelmiller, and Aviva B. Dorfman. *The Work Sampling System: Omnibus Guidelines: Preschool through Third Grade*. 3rd ed. Ann Arbor, Mich.: Rebus, 1994.

National Center for Learning Disabilities. "No Child Left Behind: Determining Appropriate Assessment Accommodations for Students with Disabilities." 2005. www.ncld.org/publications-a-more/parent -advocacy-guides/no-child-left-behind-determining-appropriate-assessment-accommodations-for -students-with-disabilities (accessed March 24, 2010).

National Council of Teachers of Mathematics (NCTM). *Principles and Standards for School Mathematics*. Reston, Va.: NCTM, 2000.

———. "Position Statement: What Is Important in Early Childhood Mathematics?" 2007. www.nctm.org /about/content.aspx?id=12590 (accessed March 24, 2010).

Newcomer, Phyllis L. *Diagnostic Achievement Battery*. 3rd ed. Austin, Tex.: Pro-Ed, 2001.

Newcomer, Phyllis L., and Herbert P. Ginsburg. *Assessment Probes for the Diagnostic Achievement Test-3*. Austin, Tex.: Pro-Ed, 2001.

Piaget, Jean. *The Child's Conception of the World*. Translated by J. A. A. Tomlinson. Totowa, N.J.: Littlefield, Adams & Co., 1976.

Wainer, Howard. *Measurement Problems* (Program Statistics Research No. 92-12). Ewing, N.J.: Educational Testing Service, 1992.

Wechsler, David. *WIAT-II: Wechsler Individual Achievement Test—Second Edition.* San Antonio, Tex.: Psychological Corp., 2001.

Woodcock, Richard W., Kevin S. McGrew, and Nancy Mather. *Woodcock–Johnson III NU Tests of Achievement.* Itasca, Ill.: Riverside Publishing, 2001.

CHAPTER 5

Number and Operations: Organizing Your Curriculum to Develop Computational Fluency

Edward C. Rathmell
and Anthony J. Gabriele

This chapter focuses on number and operations. The primary goals are to provide the following:

- ◆ Expectations for mathematical proficiency in number and operations

- ◆ Suggestions for organizing the curriculum to conceptualize and develop proficiency with number and operations in the classroom

- ◆ Suggestions for special adaptations and accommodations (based on available research on students with difficulties learning mathematics) that teachers might use to ensure that proficiency with number is a realistic goal for all students

The Number and Operations Standard

The Number and Operations Standard of the *Principles and Standards for School Mathematics* specifies that "instructional programs from prekindergarten through grade 12 should enable all students to understand numbers, ways of representing numbers, relationships among numbers, and number systems; understand meanings of operations and how they relate to one another; and compute fluently and make reasonable estimates" (National Council of Teachers of Mathematics [NCTM] 2000, p. 32).

Each dimension represents a different aspect of being mathematically proficient with number, and they are highly interrelated. From the perspective of teaching and learning numbers, figure 5.1 shows their interrelatedness. Understanding numbers and understanding the meaning of operations, which underlie the behaviors often attributed to number sense, are foundational to computing fluently (Russell 2000). This importance is not diminished in the recent *Common Core State Standards*, which includes number and operations in base ten as a critical domain for grades K–5 and Ratios and Proportional Relationships and the Number System as domains in grade 6 (Council of Chief State School Officers [CCSSO] 2010).

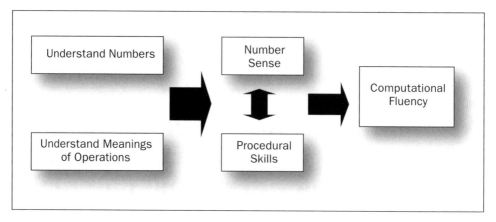

Fig. 5.1. Reconfiguration of the Number and Operations Standard: interrelationships of mathematics proficiency aspects

What is computational fluency?

As Russell (2000) described, students who are computationally fluent can compute accurately, efficiently, *and flexibly*. This description of computational fluency goes beyond typical descriptions, which focus on the accuracy and efficiency with which students compute. For example, flexibility refers to students' abilities to make smart choices about which of multiple strategies or tools would be most useful for solving a particular problem. Is a precise answer required, or would an estimate do? If the problem needs mental computation, how might the student transform, decompose, or recombine the numbers to use known number facts? This ability presupposes a disposition to think about alternative strategies and the knowledge to (1) execute a variety of strategies, (2) use different ways to represent computational situations, (3) understand when and why strategies are appropriate to use, and (4) judge reasonableness and monitor the processes of computation, as shown in figure 5.2.

By including flexibility as a defining attribute of computational fluency, Russell (2000) has helped to articulate NCTM's vision of the importance of computational facility grounded in an understanding of number and operations. Thus, for students to learn how to compute fluently, teachers must find ways of helping students develop procedural skills together with a sense of number. Aspects of number sense particularly important to developing computational fluency include understanding (1) what numbers mean and how students can use them, (2) how numbers are interrelated (including their relative size), (3) various ways to represent numbers, (4) meanings of operations on numbers (e.g., how an arithmetic operation can model real-world transformations or situations), and (5) the effects of operating on numbers (e.g., multiplying by a fraction less than 1 will result in an answer smaller than the multiplicand) (Fennell 1993).

To help illustrate the meaning of computational fluency, let us consider what students will know and be able to do to exhibit computational fluency with subtraction:

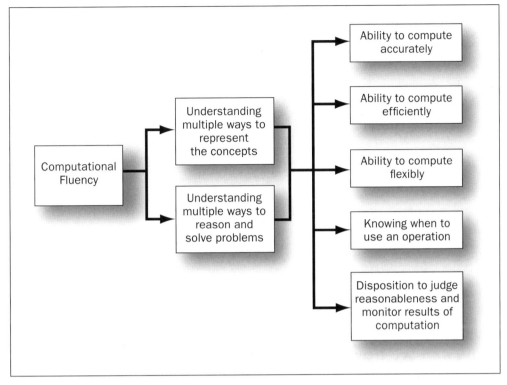

Fig. 5.2. Attributes of a computationally fluent student

◆ Efficiently and accurately compute answers to subtraction situations

◆ Recognize and use a variety of verbal, visual, and concrete representations for take-away, missing-addend, and comparison subtraction situations

◆ Understand the part–part–whole concept and, consequently, understand relationships between subtraction and addition (i.e., one can add both parts to get the whole and subtract either part from the whole to get the other part)

◆ Flexibly use a variety of thinking strategies to mentally compute answers to subtraction situations (fig. 5.3)

◆ Strategically choose appropriate computational methods, including mental calculations, invented procedures, standard algorithms, and estimates, that will lead to accurate and/or efficient results

◆ Monitor their work to ensure the accuracy and reasonableness of their results

◆ Believe that they can make sense of subtraction in and out of the mathematics classroom

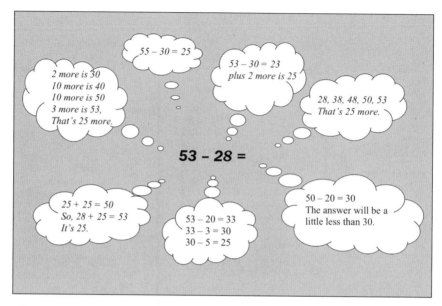

Fig. 5.3. Some computational and estimation subtraction strategies

How can we develop this computational fluency?

One thing is clear about the effects of current mathematics curricula: few students develop computational fluency as defined here.

Some argue that traditional instruction is unlikely to develop the computational fluency that this chapter describes. This is partly because instruction focusing primarily on skill acquisition, through repeated practice, appears to interfere with the kind of learning required to develop number sense (e.g., Pesk and Kirshner 2000; Ritchhart and Perkins 2000). Some students may learn to compute quickly and accurately from this instructional approach. However, these students often have neither the disposition nor the broad understandings necessary to use their skills flexibly, to judge the reasonableness of their solutions, or to monitor the results of their computation. For example, after students have learned the standard subtraction algorithm and renaming across zeros, far too many will use that same written algorithm to find the difference between 400 and 398. They do not display the strategic flexibility in their approach to computation to recognize that 400 is only two more than 398 and to know that if the parts of 400 are 398 and 2, then 400 − 398 = 2. Despite being able to use an algorithm to arrive at a correct answer, these students have not achieved proficiency with respect to Number and Operations.

Standards-based efforts generally advocate problem-centered instruction that emphasizes solving challenging problems that are open ended or that students can solve using a variety of strategies. Students are encouraged to justify their representation of the situation and their mathematical reasoning to others and to listen to their peers' explanations in an attempt to make sense of mathematical concepts and procedures. Through such integrated problem-centered learning opportunities and open-ended discussions, involving specific justification of their representations and their reasoning strategies, students are expected to develop the following:

◆ Rich understanding of the mathematical concepts underlying the solutions to problems

◆ Flexible problem-solving strategies and content-specific mathematical thinking

◆ Adaptive beliefs related to the discipline of mathematics

◆ Greater confidence in their abilities to think mathematically

Research indicates that reform-based instruction, implemented with fidelity, can result in impressive percentages of students reaching or exceeding standards of proficiency in procedural skills and conceptual understanding (e.g., Schoenfeld 2002).

As students begin to use and make sense of new representations and thinking strategies that evolve from these problem-centered experiences, they develop better computational fluency. For example, when students first learn to subtract, they commonly count all, that is, to count the whole, count the part that is subtracted, and then count the part that is left. As they develop the part–part–whole concept and a sense of number, they realize that taking away one or two is just like counting back one or two. This understanding lets them compute without using counters or their fingers.

Often, as use of new representations emerges, new thinking strategies develop. For example, representations for missing-addend and comparison situations may lead to counting or adding up as a new thinking strategy for subtraction. That new thinking lets students compute quickly and accurately when the number being subtracted is nearly the same size as the whole. Using larger numbers often provokes these new thinking strategies (fig. 5.4).

Presenting problems with different structures and situations, such as having the part that is left hidden, can provoke new thinking (table 5.1).

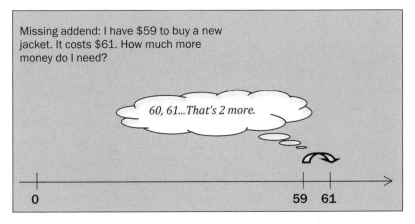

Fig. 5.4. Representations can lead to new thinking—*Continues*

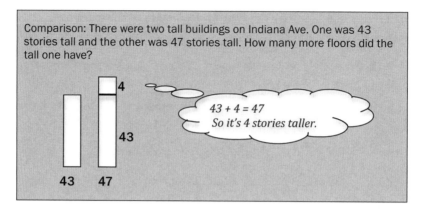

Fig. 5.4. Representations can lead to new thinking—*Continued*

Table 5.1
Different approaches to present problems and encourage new thinking

Teacher shows	Teacher asks	Common student's response
Show eight bear counters.	How many bears are there?	After counting . . . "Eight."
Make a cave with your hand and move all eight bears into the cave so students cannot see them.	How many bears are in the cave?	"Eight."
Move six bears out of the cave all at the same time—not one at a time.	How many bears came out of the cave?	After counting . . . "Six."
Keep the remaining two bears out of sight—still in the cave.	How many bears are still in the cave?	After inventing a solution . . . "Two."
Remove your hand and show the two bears that were still in the cave.	That's right, there are still two bears in the cave. How did you know that?	Typical explanations— "I counted . . . seven, eight. That's two more." "I know that 6 + 2 = 8, so it's 2." "I put up eight fingers and took six away. There are two left."
	Let's talk about how each of those strategies works. . . .	

Hiding counters provokes students to invent new strategies. They cannot count the part that is left because it is still covered. Development of the part–part–whole concept together with this structured use of models often leads to using known addition facts to help students solve subtraction problems, that is, a new way of thinking.

Recognizing that they can use known addition facts to solve subtraction problems is a significant step forward in students' development of computational fluency with subtraction. These students now begin to integrate their knowledge of addition and subtraction. Before this, these topics were two separate entities for many students.

As students become more accurate, confident, and efficient using a variety of counting and computation strategies, over time they typically begin to master (i.e., can retrieve from memory) the basic number combinations or facts. Furthermore, learning to use these different computation strategies to solve basic fact problems lets students begin to extend using these different representations and thinking strategies so they can mentally compute and estimate with larger numbers. Figure 5.3 shows different thinking strategies that students who are computationally fluent with subtraction exhibit.

Challenges for students with learning difficulties

The standards and focus statements related to number and operations—from the *Principles and Standards* (NCTM 2000) Number and Operations Standard, *Curriculum Focal Points for Prekindergarten through Grade 8 Mathematics: A Quest for Coherence* (NCTM 2006), and *Common Core State Standards* (CCSSO 2010)—all set ambitious targets for classroom instruction. As the Equity Principle of *Principles and Standards* clearly states, all students are assumed to be capable of reaching these high expectations given the opportunity, with good instruction and strong support in the form of extra individual and/or small-group attention. Researchers have questioned whether the *Principles and Standards* and the *Curriculum Focal Points* are appropriate for students who have difficulties learning mathematics (e.g., Bottge et al. 2007; Vaughn, Klinger, and Hughes 2000). Many students with learning problems do not perform like other students. The debate among mathematics educators and special educators is how to respond to this situation instructionally.

One view is that many of the difficulties students have are a function of approaches to teaching mathematics that are limited and that many, if not all, of the associated difficulties diminish with thoughtfully implemented focused instruction. Another view, generally of special educators, is that these individual differences in performance reflect real deficits that require individualized instructional adaptations or accommodations.

Although research evidence indicates that standards-based instruction can lead to the kind of mathematical outcomes that *Principles and Standards* articulates, little of this research has specifically examined how students with mathematical difficulties learn from standards-based instruction in the regular classroom (Bottge et al. 2007). The existing research has yielded mixed results (e.g., Baxter et al. 2002; Baxter, Woodward, and Olson 2001; Bottge et al. 2007; Carnine, Jones, and Dixon 1994; Kroesbergen and van Luit 2002; Kroesbergen, van Luit, and Maas 2004; Woodward and Baxter 1997). Therefore, we cannot now refer directly to empirical research to make informed decisions about whether standards-based instruction will pose special challenges or enhanced learning opportunities for students with difficulties learning mathematics. Instead, we present a brief review of relevant research on the numerical competencies of students with learning difficulties. Based on this review, a discussion follows about how regular classrooms that implement standards-based instruction might change to ensure the success of these students.

In general, research indicates that students with mathematics learning difficulties (MD students) are less proficient with numbers and operations than non-MD students as early as kindergarten. Through the elementary years, many of these numerical competencies appear to improve, suggesting that these early performance differences are more likely to represent developmental delays than cognitive deficits (Geary et al. 2004). However, as mentioned, one must interpret this and other general patterns cautiously, in part because of differences in how researchers have identified learning difficulties in mathematics (Mazzocco 2007). As we describe in the following, recent research suggests that subtypes of MD children exist that present different profiles of performance.

Number comprehension skills, which include among other things a child's ability to associate and translate among number names (e.g., three), symbolic representations (e.g., 3), and associated quantities (e.g., $\mathscr{O}\mathscr{O}\mathscr{O}$), as well as understand ordering among these quantities (e.g., 3 > 2), appears to be generally intact in MD children, at least for single-digit numbers (Geary and Hoard 2001). However, some research indicates that children who display disabilities in both reading and mathematics (RD/MD) show developmental delays in number comprehension and skills in relation to children without MD and those children who display difficulties only with mathematics (Geary, Hamson, and Hoard 2000).

With respect to counting knowledge, research by Geary, Hamson, and Hoard (2000) suggests that children with MD/RD and MD in first and second grade understand most of the basic counting principles, such as one-to-one correspondence, stable order (the order of the number name must be invariant across sets), and cardinality. However, they do not appear to understand the order-irrelevance principle: that counting a set of objects from left to right yields the same number as counting the same set from right to left. Instead, many such children appear to believe that counting must proceed in succession without skipping around. This misunderstanding may in part explain rigidity in their counting behavior in the early grades and why in this and similar studies children with MD use sophisticated counting strategies such as "count on" less often than typically achieving children in first grade.

With respect to strategies to solve addition and subtraction problems, as well as their more frequent use of less sophisticated counting strategies such as "count all," longitudinal research indicates that in first grade children with MD make more counting errors and tend to rely more heavily on finger counting than children without MD (Geary, Hamson, and Hoard 2000; Geary et al. 2004). By second or third grade, children with MD are just as accurate as typically achieving students in using counting strategies on single-digit computation, making significant progress in both their use of sophisticated counting strategies and their reduction of errors. However, they continue to rely more heavily on their use of fingers than typically achieving students, with differences found as late as fifth grade (Geary et al. 2004).

Other research indicates that children with MD are less flexible and slower in using computation strategies than typically achieving students. In the study of Verschaffel and colleagues (2007), students were briefly taught two different derived-fact strategies (i.e., the make-ten and near-doubles strategies) and then given a set of computation items and allowed to choose the strategy they preferred to solve the problems. Two-thirds of the typically achieving children used both strategies when solving the computation items; in contrast, less than one-third of the children with MD used both strategies. Instead, most children with MD relied on one strategy (usually the make-ten strategy) to solve all the computation problems.

With respect to more complex arithmetic computation, Russell and Ginsburg (1984) found that fourth-grade children with MD committed more errors than did their IQ-matched, normal-achieving peers. Errors included both misalignment of numbers while writing down partial answers and regrouping errors. These errors appear to be due to difficulty in monitoring and self-correcting errors rather than poor conceptual understanding of the base-ten system.

Finally, a consistent research finding is the relative difficulty that children with MD have retrieving basic number combinations from memory (e.g., Fleischner, Garnett, and Shepherd 1982; Goldman, Pellegrino, and Mertz 1988). For example, when MD students must respond quickly (within three seconds) to simple arithmetic problems (thus precluding their use of counting or other computational strategies), they are far more likely to make errors or to not answer than typically achieving children (Jordan and Montani 1997). Some speculate that the difficulties that children with MD have mastering basic number combinations may be the result of poor working-memory resources and overreliance on inefficient counting procedures (which take longer to execute), which compound these students' difficulties forming memory associations and ultimately basic number combinations in long-term memory (Geary et al. 2007).

In summary, research on the numerical competencies of students with learning difficulties has demonstrated the following weaknesses that need to be addressed. Such students rely on less efficient counting strategies and use their fingers for extended periods, are more rigid in their use of a particular strategy once they have learned to use it accurately, are less flexible in their use of a variety of strategies, and have difficulties mastering basic number combinations.

These results imply that MD students need special instructional attention to develop and use computational strategies that will help them learn the basic facts and solve more complex arithmetic problems. We believe that flexibility in the use of strategies can be developed, but these students need much more time and focused attention engaging in opportunities to make strategic choices.

Reorganizing the Curriculum to Help All Students Develop Computational Fluency

With this broad-based view of computational fluency, students need to make sense of multiple representations and multiple ways of reasoning with numbers and operations. That takes time, much more time than typical textbooks currently allot to any topic. For example, a typical second- or third-grade textbook has two or three units of instruction involving subtraction. Even if a teacher spends three weeks with each unit, the students have only nine weeks to make sense of that topic. Skills may be developed. Understandings may grow. But not enough time will be available for students to develop the number sense and flexibility intended in the NCTM vision.

Computational fluency requires both number sense and proficiency with procedural skills. The complexity of developing number sense and flexibility demands that instructional activities be distributed throughout the school year. Students simply cannot develop all these understandings and skills in a few relatively brief units of instruction. They must have opportunities to invent and discuss solutions to problems and the time to make sense of multiple ways to represent the ideas and multiple strategies to solve a variety of problems. Flexible thinking is crucial at this stage. In fact, these experiences help students build a basis for making strategic choices in using mental computation and estimation strategies.

The following discussion elaborates on a reorganization of the curriculum designed to accomplish both the development of number sense and fluency with procedural skills. It is based on distributing experiences in solving word problems throughout the school year, yet giving special attention to the thinking students need in order to make sense of standard algorithms. Here is an overview of organizing the curriculum for computational fluency:

Distribute Instruction for Number Sense

Provide brief exploratory conceptual activities and experiences solving word problems nearly every school day to help students make sense of and develop strategic choice with multiple representations and thinking strategies related to an important topic and to develop a disposition to make strategic choices.

Special Attention to Computational Skills

If there are expectations of fluency with any computational skills, those skills need special attention. Daily two- to three-minute conceptual previews can be used for about two weeks before beginning the formal unit of instruction for a skill. Similarly, brief practice, strategic choice, and connections activities can be continued for several weeks after the formal unit of instruction.

Basic Facts

First Stage: Conceptual Previews

Structure word problems to encourage or provoke targeted thinking, and then illustrate concretely after students explain their thinking.

Second Stage: Practice and Strategic Choice

Continue solving word problems with an emphasis on making strategic choices of thinking strategies to develop flexibility of their use.

Symbolic Skills

First Stage: Conceptual Previews

Concretely preview the thinking students will use with the standard algorithm by asking them to solve and explain one computational situation with manipulatives or diagrams (symbols may interfere at this stage).

Second Stage: Record the Symbolic Procedure

Connect each digit that is written in an algorithm, step by step, to the actions students have used to solve the computations concretely.

Third Stage: Practice and Connections

Practice one or two problems each day, together with conceptual explanations that continue to connect the symbolic procedure to the actions on manipulatives or diagrams, and monitor results.

Continue to Develop Number Sense

Standard algorithms need to be understood as only one of many different ways for students to compute. Continued emphasis on making strategic choices of thinking strategies, after teaching a standard algorithm, enhances the development of number sense and flexible use of number and operations.

Number sense

Mathematics educators have long argued that students with a deep understanding of a mathematical topic are much more likely to retain and apply mathematical ideas flexibly. Most students need many opportunities to deeply understand a topic. For MD students, who often exhibit difficulties retaining and applying their mathematical ideas flexibly, furnishing more conceptual experiences than usual is even more crucial. These conceptual experiences are the very ones that let children develop number sense and flexibility, and instruction should focus on those goals. Therefore, an important step in proficiency in number and operations is to organize the curriculum to give children many more opportunities to make sense of these mathematical ideas—not just to practice skills but to conceptually understand.

Also, as students have multiple opportunities to explore different ways to represent and solve these word problems, they begin to believe that they can make sense of mathematics. As they try new ideas and confirm their thinking, they begin to take ownership of their learning and develop confidence. As they reflect on when specific reasoning is and is not efficient, they begin to develop strategic approaches to solve and monitor their solutions to problems.

Distributed instruction, which presents mathematical topics repeatedly throughout the academic year rather than in one chapter or unit, is one example of such a reorganization (Resnick et al. 1991). By using five-minute instructional tasks related to a key mathematical topic almost daily throughout the school year, distributed instruction offers many additional opportunities for conceptual experiences, that is, for students to make sense of the topic. Instructional activities, distributed throughout the school year, also give opportunities to devote more time to a mathematical idea. Teachers later recapture the time devoted to conceptual understandings by the reduced need both for practice and for repeating instruction in later years. These brief activities can supplement the regular mathematics curriculum. They will not always have direct connections to the regular mathematics lessons for the day; they are ongoing supplements to help students develop number sense and flexibility for some key topic or focal point. Suppose that a teacher presents a unit on geometry in the regular curriculum. If the teacher is helping students develop number sense related to addition and subtraction, these daily five-minute supplements will focus on addition and subtraction and will not necessarily be connected to geometry.

The psychology-of-learning literature documents distributed learning's positive effects on learning and retention (Rohrer and Taylor [2006]; see Cepeda et al. [2006] for a review). One fifty-minute practice session is not nearly as effective as five ten-minute practice sessions distributed throughout a week. Similarly, distributing conceptual experiences throughout the school year may be more effective than including them all in a two- or three-week unit of instruction. These brief, near-daily experiences, when supplementing the regular curriculum, actually give students more time to make sense of a topic as well as offer the benefit of spacing learning opportunities.

The following vignettes describe the classroom experiences of several teachers who have reorganized their curriculum to afford additional conceptual, problem-solving, and skill development experiences. In each case, we highlight how this reorganization affected a particular student with learning difficulties.

Proportional reasoning vignette, grade 8

The following vignette explores proportional reasoning.

Background. Tori is an eighth-grade student who has experienced difficulties with reading comprehension and mathematical concepts throughout elementary school. She participated in a Title I program for much of that time. Because of her low rate of success, she was not well motivated to even attempt school assignments. In particular, her perseverance with mathematics tasks was quite low. Tori had no individualized education program (IEP).

Because of a reassignment of teaching duties, Mrs. Shindelar had Tori as a student in grade 3 and then again in grades 7 and 8 as her mathematics teacher. Mrs. Shindelar expected that Tori would have difficulty with the eighth-grade mathematics curriculum because of her learning difficulties in grade 7, her deficiencies in number sense, and her inconsistencies with motivation. Furthermore, because the eighth-grade mathematics curriculum emphasizes problem solving with proportional reasoning, Mrs. Shindelar was afraid that Tori would get left behind, get frustrated, and perhaps quit trying altogether.

Accommodations. Mrs. Shindelar made two accommodations that proved to be a great benefit to Tori. First, after many invitations, Tori began coming to Mrs. Shindelar's classroom for a few minutes to discuss her homework before her mathematics class. Doing so let Tori get feedback about her homework, but more important, gave them both a time to model the problems and discuss the homework conceptually. Tori realized that these teacher–student interactions were helpful and began to take advantage of the extra time regularly. Not only did it give her an opportunity to correct errors on her homework assignment, but the modeling and discussion about the concepts also helped her make sense of the mathematics.

Mrs. Shindelar made a second accommodation in the mathematics curriculum for the entire class, not just Tori. Knowing that solving problems with proportional reasoning was a key topic for grade 8, Mrs. Shindelar decided to use a problem or conceptual task each day for about a month before teaching an instructional unit on proportions. These conceptual activities were simply introductory problems involving proportions.

The students were expected to work on their own, briefly, and to try to solve the problem by using more than one strategy. Then the students paired up and discussed their solution strategies. They shared a few of these strategies with the entire class. After illustrating a few different strategies on the chalkboard, Mrs. Shindelar led the students in a discussion about the strategies and relationships among them. These activities took about ten minutes each day. This extra time to think about proportions gave all the students, including Tori, a much greater opportunity to make sense of different solution strategies for proportional reasoning problems. Instead of two weeks during the instructional unit, the students had six weeks (albeit distributed into relatively brief conceptual experiences) to develop understanding.

Activities. "I recall vividly some of the things that Tori did while she was working on these problems," Mrs. Shindelar recalls. "The first task was 'Sixteen doughnuts cost $8.00. How much would 12 doughnuts cost?' Tori just sat there with her hands folded on her desktop, her paper blank, not knowing what to do. The student that shared with Tori explained how she divided to find the cost of one doughnut, then multiplied by 12. This made sense to Tori. In fact, she used that same thinking process, a unitizing approach, to solve each of the next few problems."

A few days after the first task, Mrs. Shindelar gave the following problem: "If 5 chocolates cost 75¢, how much will 13 chocolates cost?" Tori continued to use a unitizing approach to solve this problem, just as she and some other students had done for the past few days. However, another student's thinking launched the class into a new approach, the ratio table. This student reasoned that "If 5 chocolates cost 75¢, then 10 would cost twice as much, or $1.50. So, 15 would cost $0.75 plus $1.50, or $2.25. Then, since 1 chocolate would cost 15¢, you could just subtract 15¢ twice from the $2.25 to get the cost for 13 chocolates. So, $2.25 minus $0.30 is just $1.95."

This encounter let Mrs. Shindelar organize this thinking by using a ratio table. She wrote the following and explained: Doubling the 5 and 75 gets the next entry. Adding the 5 and 10 is 15 chocolates, so adding the corresponding costs is the cost for 15, or $2.25.

Dividing $1.50 by 10 gives the cost for one chocolate. Then subtracting the cost of one chocolate twice gets the entry for 13 chocolates, namely, $1.95. See figure 5.5.

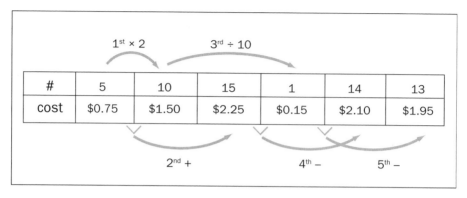

Fig. 5.5. Using a ratio table to solve a proportion problem

Many students used the ratio table for the rest of the conceptual tasks before the instructional unit on proportions. Some students began to look for efficient ways to combine the information in the tables to get the result. They discussed these methods daily.

Tori used the ratio table to organize her work; however, she continued to unitize the problem and repeatedly add to find her solution. See figure 5.6.

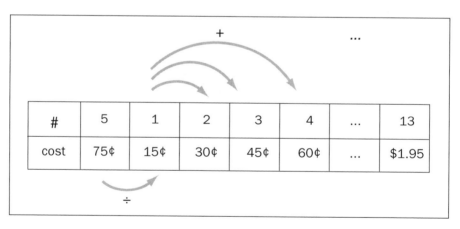

Fig. 5.6. How Tori used a ratio table to solve proportion problems

Although Tori had made progress during the distributed daily conceptual experiences that preceded the instructional unit of proportions, she still tended to rely on less sophisticated reasoning and computational strategies. This situation posed another challenge for Mrs. Shindelar: when and how to provoke Tori to use more sophisticated strategies.

> It was hard for me to decide when to push a bit to get Tori to use the ratio table more
> efficiently. I often reminded myself that this was really unitizing for her and she needed
> more time to feel comfortable enough to eliminate some of the additive steps. Still, this

was progress. Tori had no strategy when we started. Now she could successfully represent and answer the problems. As we moved into the instructional unit on proportions, I tried to bring Tori into class discussions. I accepted her explanations with elongated tables. But I tried to ask reflective questions: Why does this table have fewer steps? What is the first (second) entry in this table? Why do you think this student chose to use this approach? Why do you think this student started with this operation to get the second entry?

Breakthrough. The first real breakthrough came when the teacher asked Tori to reflect on and describe other students' work. It created a situation where she had to think about why another student made the next entry in the table. She gradually learned to recognize these patterns and became more confident about describing other students' methods. Doing so led her to try some of them, with success.

Summary. Tori still struggled with mathematics, but she gradually learned some more efficient ways to use a ratio table to help solve proportional reasoning problems. It gave her a tool that she used to solve problems for the rest of the school year. Doing so was not easy for her. Making sense of this approach took her much longer than for other students in the class. However, because Tori had more than six weeks to understand proportional thinking rather than the two weeks allotted for the instructional unit, she got the extra time she needed to eventually solve proportion problems.

Mrs. Shindelar encouraged Tori to take advantage of extra time to discuss her homework and her lessons. And Mrs. Shindelar reorganized the curriculum to create more conceptual experiences to help all her students, including Tori, make better sense of proportions and how to solve related problems. The ratio table helped Tori learn a new way to represent proportional situations. The ratio table made more efficient ways of thinking accessible to Tori. Then, Mrs. Shindelar used the ratio table to generate discussions, which provoked Tori to make sense of more efficient reasoning strategies.

The word problems to help students develop their number sense can be brief, but frequent, five-minute activities. To give students more time to make sense, teachers should also present more complex tasks, with more time for exploration and discussion of solutions, as well. But time for more complex problems will not be available each day.

Special attention to computational skills

Problem-solving experiences, as the preceding vignette described, help students make sense of mathematics and to develop number sense and flexibility. But even with these ongoing experiences, some students may not develop the specific thinking that will help them become proficient with basic facts and/or mental mathematics.

Stage one: Conceptual previews for basic facts

Students, especially students with learning difficulties, often continue to use inefficient counting strategies with the harder basic facts. Daily previews structured to encourage, or even provoke, the development of a specific thinking strategy are helpful to them.

These learning experiences have a different purpose from those designed to promote number sense. Although flexibility in using reasoning strategies is an overall goal, a more structured learning experience may be needed to help students become proficient with basic facts. Letting students continue to choose their own inefficient reasoning strategy probably will not help them learn the harder basic facts. Provoking the students to learn an efficient reasoning strategy may be needed.

For example, some students may not have learned to split problems into parts they know that can be added to answer the harder multiplication facts. With a few brief and structured daily previews, including folding graph paper into two parts so they can add the parts, these students will make sense of thinking 6 × 8 is just twice as much as 3 eights: 24 + 24. Or, 6 × 8 is just 8 more than 5 eights: 40 + 8 (fig. 5.7).

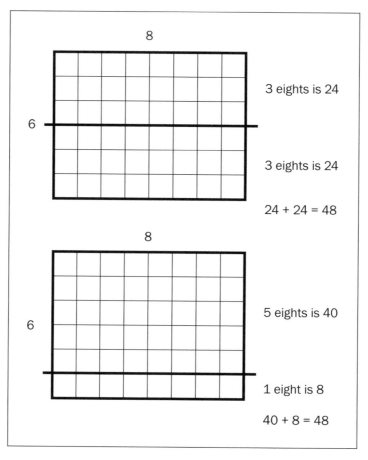

Fig. 5.7. Using folded graph paper to grasp multiplication

In the following vignettes, each teacher gave the students brief daily activities to help them with specific reasoning strategies. The teachers illustrated the ideas concretely, when appropriate. The activities were structured to elicit and often to provoke the specific thinking that an upcoming skill needed. Mrs. Louk helped Jodi learn a thinking strategy to split a multiplication problem into parts and then add the parts. This approach let her use facts that she knew to help her solve facts that she did not yet know.

Multiplication facts and algorithm vignette, grade 5

Jodi's IEP indicated that she knew the basic multiplication facts, but Mrs. Louk found that Jodi had difficulties learning the multiplication algorithm because she did not know the "hard" multiplication facts well. By giving Jodi extra time to concretely learn new thinking strategies for these basic multiplication facts, Mrs. Louk helped Jodi develop confidence that she could figure out the basic facts she did not know. With her new abilities to figure out the hard multiplication facts and with the additional support of a multiplication table, she could learn the multiplication algorithm.

Background. Jodi received special education services for reading and language, with additional support for mathematics. According to her IEP at the end of grade 4, she had mastered basic multiplication facts but needed to learn the multiplication algorithm. Her IEP indicated that she needed to have directions explained individually, have help when reading difficult passages, have help editing written work and correcting mistakes, have shortened assignments, do some written assignments orally, have tests read aloud, and attend a support room to receive more help as needed.

In grade 5, Jodi integrated into the regular classroom for all subjects except reading and language arts. She continued to struggle with most of the harder multiplication facts, those involving fours, sixes, sevens, eights, and nines. She was unsuccessful with the multiplication algorithm.

Furthermore, Jodi was quick to demonstrate frustration with tasks, and her temper flared at slight provocations. Mrs. Louk needed to always be aware of Jodi's emotional state. At the first indication of frustration, Mrs. Louk needed to be there to help remedy the situation before Jodi was out of control. Jodi needed to feel included and accepted. She was far more aware of the social standing of other students than were any of her classmates. She could not work in a group with one of the other students, despite continued discussions about appropriate classroom behavior and the need to show respect to others.

Accommodations. It quickly became apparent that Jodi benefited from using concrete materials to help her make sense of mathematics. Since she had not mastered multiplication facts, she was allowed to use a multiplication table whenever the task involved problem solving or went beyond the basic facts. Because her reading skills were low, Mrs. Louk read many problems to her. And because of Jodi's erratic emotional states, Mrs. Louk often stood by her to offer support and encouragement and to maintain awareness of her anxieties. Furthermore, Jodi never had a time limit when completing an assessment task.

Mrs. Louk's school used a DEAR (Drop Everything and Read) program for twenty minutes each day. She took a few minutes from this time each day to help Jodi, sometimes individually and sometimes with a few other students who needed help with multiplication facts.

This application of distributed instruction let Jodi concretely represent actions, which enabled her to learn a new reasoning strategy that helped her easily figure out the hard multiplication facts.

Activities. Because Jodi had difficulty memorizing the harder multiplication facts, Mrs. Louk began a daily routine of using manipulatives to help develop efficient reasoning strategies. For example, by folding a piece of graph paper that is four by six, Jodi can see that 4 sixes is twice as much as 2 sixes. This thinking helped Jodi make sense of using doubles to help her with the harder facts. She could just think of half as much and then double it.

As the class studied the multiplication algorithms for whole numbers and decimals, Jodi was allowed to use a multiplication table as needed. This accommodation, together with her increased confidence with the basic facts, enabled her to learn the procedures. She even began the assignments without Mrs. Louk's jump-starting her. She still needed encouragement and often asked, "Did I do this right?"

Breakthrough. Mrs. Louk was thrilled one day when Jodi called her over and then said, "Never mind, I just forgot the three." Jodi actually looked for and found her own mistake. On the test at the end of the unit for multiplying decimals, she missed only two items, both involving multiplication with money. She solved the other computation problems.

Summary. Jodi continues to need support and encouragement. She continues to get frustrated. She still has emotional problems and needs constant attention. However, she has learned the multiplication algorithms for whole numbers and decimals. And she has made significant progress with the basic facts. More important, Jodi is now willing to try new tasks and does not shut down nearly as soon when she does not succeed immediately.

Mrs. Louk tried to help Jodi make sense of mathematics. She gave a few minutes of additional conceptual experiences each day. This extra time, support, and encouragement are helping Jodi make progress. Because she is making sense of the multiplication algorithm, she can feel successful and included. Jodi still has problems, but she now can do the same work as the rest of the class.

Stage two: Practice and strategic choice for basic facts

Even after students have made sense of new thinking, some practice with explanations helps them develop basic fact retrieval and confidence in their ability to use numbers fluently. In the following vignette, Mrs. Cotton gave Jeremy previews to help him learn to use counting back and counting up, but then she continued to ask him to explain how to solve fact problems with that new thinking. That practice with explanations helped him develop both skills and confidence. These thinking strategies and the practice with explanations gave him a way to easily solve many addition and subtraction facts and understand the part–part–whole concept.

Addition and subtraction facts vignette, grade 1

As a first-grade teacher, Mrs. Cotton used daily word problems to help her students begin to make sense of addition and subtraction. Initially, in her regular mathematics curriculum,

Mrs. Cotton wanted to make sure that the students could represent addition and subtraction situations with counters and begin to learn efficient counting strategies.

Background. As Jeremy entered grade 1, it was immediately obvious that he would need special attention. He was easily distracted. He had difficulty focusing on a task. He had little confidence. He would not try anything on his own. He liked to have something in his hands at all times. His writing was sloppy. He made many symbol reversals. He had difficulty articulating his thinking, both orally and in writing. He did not know the meaning of addition or subtraction. And he could not make connections between what he knew and something new, even if the tasks differed only slightly. Jeremy had no IEP.

Early in the school year, Jeremy would just shut down during mathematics activities. He would play with things in his supply tub, mess with his shoes, twist himself into a pretzel in his chair, or talk and bother other students at his table. He did not show interest in mathematics and did not see a need to attend to the tasks. His frustration and anxiety were obvious anytime someone asked him to respond.

As students came into Mrs. Cotton's room each morning, they had about fifteen minutes to put their things away, fill their water bottles, mark their choice for lunch, organize their desks, and return library books before classes officially began. Then the students had another few minutes to individually write an equation for the number of the day. Mrs. Cotton often used this time to briefly meet with individual students.

Accommodations. Mrs. Cotton used this early-morning routine to help Jeremy. Two or three times each week, she met with him to give five to ten minutes of instruction. During this time she reviewed previous work, previewed ideas to help prepare Jeremy for the mathematics lesson that day, and sometimes previewed new concepts that she was not going to introduce to the rest of the class until two or three weeks later. She also informed Jeremy's parents about these mathematics activities, supplied them with appropriate materials, and asked them to help Jeremy for five to ten minutes a few nights each week.

At this time as part of her mathematics instruction, Mrs. Cotton was using a five-minute daily routine to help the children make sense of addition and subtraction situations. She expected the class to mentally solve, and then explain their solutions to, word problems such as "James had 6 pencils. Four of them were red. The rest were blue. How many were blue?" Most children began the year counting from 1 to solve similar problems. However, as some children shared counting-up and counting-back strategies that they used, other children in her class began to integrate these strategies into their own problem-solving strategies.

Activities. "We spent a lot of time at the beginning of the year learning about partitions of numbers and the part–part–whole concept," Mrs. Cotton remembers. "One of the activities that we did together involved two-sided color counters. Jeremy would grab a handful of counters and sort them into two parts, red and white. Then he would put the part that was red with the part that was white to make the whole. Next he would tell and record number stories about the counters. In another activity, I would say, 'I have six pennies altogether. I have four pennies in my right hand. How many are in my left hand?' We took turns holding the pennies and determining how many pennies were in the other hand.

"I chose these particular activities because Jeremy was a very distractible child. If he had something in his hands, it seemed to help him stay focused. During these activities, Jeremy found two ways in which he could find the missing parts. He was able to make sense of counting up and counting back as reasoning strategies to help solve problems. As he gained confidence, he began to apply counting on and counting back to the daily word problems and mental computation problems that we were doing in the mathematics class."

Breakthrough. "After about a month of this accommodation, he first participated in the mental mathematics problems that the rest of the class was solving. I was shocked the first time he raised his hand and volunteered." Here is what happened.

Mrs. Cotton: Laura had eight pennies. She put two of them in one hand. How many were in her other hand?

Jeremy: Six.

Mrs. Cotton: Great! How did you know that?

Jeremy: Well, I knew that she had two in one hand, so I just counted up until I got to eight, and that would be six more.

Mrs. Cotton: Super! What parts of eight did we find today? Can you write a number sentence to show your thinking?

Jeremy: [*He wrote "2 + 6 = 8" on the chalkboard.*]

Mrs. Cotton: You are exactly right. Now let's all use Jeremy's thinking to solve another problem.

"Jeremy was thrilled that we used his thinking to solve another problem." After this event, Jeremy began to share his thinking regularly with the whole class and during group work.

Summary. "I was really amazed at what this five to ten minutes extra a day did for Jeremy's thinking and confidence. It made a huge difference. In fact, as he began to demonstrate that he understood, I began providing less and less support, and I expected more from him. I began including other students in our extra instructional sessions. As the year progressed, Jeremy did not need that extra support nearly as often. In fact, he became one of the stronger mental problem solvers in the class."

Jeremy still needed support. He still benefited from having manipulatives in his hands. Sometimes he still got distracted or became frustrated. But his confidence soared. He participated in class activities and discussions. He explained his thinking about problems. Regularly spending a few extra minutes of conceptual activities with Jeremy gave him the extra instructional time he needed to make sense of counting on and counting back to add and subtract. Distributed instruction did not eliminate all Jeremy's problems, but it did enable him to perform addition and subtraction skills as well as other first graders.

Stage one: Concrete previews for symbolic skills

In Mrs. Shindelar's class with a focus on proportional thinking, Tori made great strides. She learned how to use a new representation, namely, the ratio table, and new reasoning strategies that were much more efficient than her initial efforts. However, just as important, these experiences were not enough for Tori to develop the fluency with written skills to let her solve proportions with symbolic procedures quickly and efficiently. She still needed something more.

Repeated opportunities to solve problems, to listen to other solution strategies, and to observe the representations that other students used are enough for many students to achieve the desired number sense and, eventually, computational fluency. But students with learning problems often need more than this. Just listening to other students and watching them use different approaches may not be sufficient for them to take ownership of those new ideas. In fact, they commonly learn a counting procedure to solve problems and then continue to use that same procedure, sometimes for several years. Unfortunately, the expectations for them to solve problems with more complexity and larger numbers grow beyond the efficient use of their counting strategies. Does distributed instruction help students with learning problems? Yes. But is it sufficient for them to develop the needed computational fluency? Not necessarily.

Teachers have long recognized that developing fluency for a written skill requires specific instructional attention, together with at least minimal practice of the skill. Consequently, at some point during the school year, teachers devote lessons to help students learn to record and practice a skill. However, if students simply memorize the procedures for a written skill, they will not have developed the understandings necessary to retain and apply the skill efficiently, meaningfully, and successfully. To understand, they must also have made sense of a reasoning strategy for use with that skill. This common thinking lets the students make connections between the step-by-step actions that they can use to solve the problem concretely and the step-by-step reasoning related to those particular symbolic procedures. Practice alone is not enough to develop these understandings. Students also need to make sense of the connections between the step-by-step actions and the step-by-step reasoning.

Students who have been using inefficient reasoning strategies to solve problems before the instructional focus on a specific written skill often concentrate on memorizing the symbolic procedure to correctly answer computation problems. These students need specific learning experiences to help them make sense of the thinking required to understand the written skill—especially important for students with learning problems in mathematics. They often continue to use inefficient strategies for extended periods, and unfortunately these inefficient strategies usually do not match the step-by-step reasoning used to record the symbols.

Distributing tasks to provoke the specific thinking needed to make sense of a skill is also extremely beneficial to students, especially to students with learning problems in mathematics. However, since the focus is now on only one specific way of thinking, one brief, concrete activity each day for two or three weeks prior to beginning the instructional unit is usually sufficient.

These activities involve using concrete models (without symbol manipulation) to solve computation problems in a specific way. The actions performed on these concrete materials and the explanations must correspond to the thinking needed to make sense of the specific written skill. Structured activities with the materials should provoke that thinking. That thinking will enable students to make sense of the written skill. Clear, logical, and confident explanations of the actions performed on materials to solve computation problems, which correspond directly to the thinking that students will need for the written skill, are convincing evidence that a student is prepared to make sense of the written skill.

In the following vignettes, the teachers made accommodations to help students learn written algorithms. In each case, teachers began by helping students make sense of the thinking before they practiced with symbols. Mrs. Louk helped Jodi make sense of fractions with a fraction–percent bar. Mrs. Pottebaum helped Jesse make sense of adding up to develop a subtraction algorithm that was meaningful to him. He did not learn the standard subtraction algorithm, but he learned one that made sense to him and he developed proficiency in using his algorithm to solve subtraction problems.

Rational number vignette, grade 5

Mrs. Louk also needed to help Jodi understand fractions, decimals, and percents and their equivalents for some common benchmarks. She recognized that Jodi would need concrete representations to make sense of these relationships. Mrs. Louk used distributed instruction in the form of an extended replacement unit to help give Jodi the time to make sense of these rational numbers.

Accommodations. Mrs. Louk asked the students to construct their own charts of fraction and percent equivalents for proper fractions with denominators of 2, 3, 4, 5, 6, 8, and 10. Jodi did not know how to begin. With guidance, she split a ten-by-ten grid into halves, fourths, and eighths. By counting the number of small squares in each part, she could write the percent to go with each fraction. After a discussion about the number of rows, she could split the grid into fifths and tenths. Again, by counting she could write the equivalent percents. Jodi could not make thirds until Mrs. Louk suggested counting three small squares at a time and shading one of them each time. It took a discussion about how to deal with the last square, but she shaded one-third of the ten-by-ten grid and wrote the percent.

Breakthrough. During this unit Jodi created several tools such as the fraction–percent charts just described. One was a percent–fraction equivalence strip (fig. 5.8). Jodi wrote some benchmark fractions, 0, $\frac{1}{4}$, $\frac{1}{2}$, $\frac{3}{4}$, and 1, on the strip. Then she wrote the percent equivalent for each. Percents that included a fraction, such as $33\frac{1}{3}\%$, stumped her. But with help from another student, she found all the fraction–percent equivalents. The experience of creating her own tool to help her relate fractions, decimals, and percents gave her confidence to complete most rational number tasks.

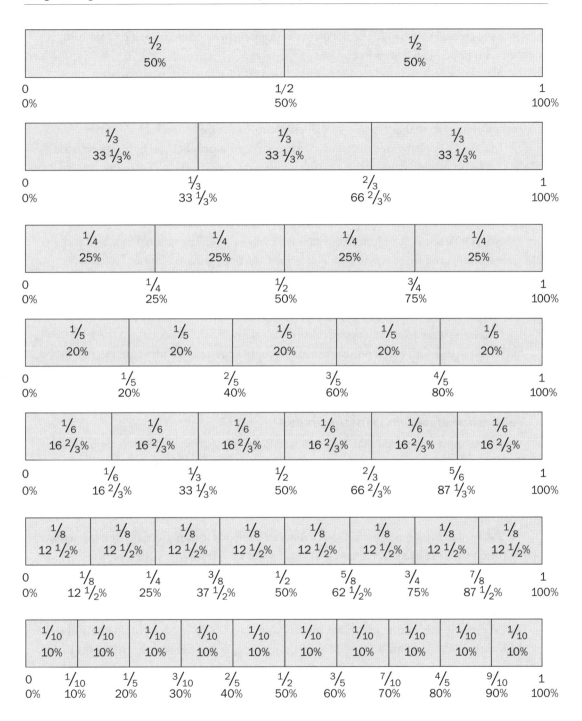

Fig. 5.8. Fraction–percent charts

Summary. Jodi was still not as proficient as most of her classmates. However, with the help of her fraction–decimal–percent tools, she could complete tasks involving rational numbers. Jodi kept these tools in her desk and used them often the rest of the school year.

Since she created these tools, she felt ownership, giving her a positive feeling about using them. They also helped her succeed.

Mrs. Louk feels that Jodi could not have used rational numbers without many concrete experiences similar to those described here. Perhaps she could memorize some equivalents, but she could not have used them in problems. Her tools enabled her to solve problems involving rational numbers—not all problems, but many of them.

Mrs. Louk implemented distributed instruction for multiplication by assigning extra daily work on making sense of the multiplication facts. Activities such as folding rectangular pieces of graph paper helped Jodi learn to use multiplication facts that she knew to figure out harder ones that she did not know yet. For the rational numbers, Mrs. Louk used a replacement unit, which extends the time devoted to the topic.

Mrs. Louk also helped Jodi construct a new way to represent rational numbers and relate fractions and percents. This tool let her use fractions and percents and solve many problems that other students in the class were expected to solve. Without it, she would not have had such success. These previews for the skills each helped a student learn a new way of thinking. That thinking, in turn, helped the students make sense of a skill.

Success also gave the students more confidence. Their confidence helped these students believe that they could make sense of mathematics. Their confidence helped them take more initiative to begin problems on their own. Their confidence helped them become students who participated and were successful.

Subtraction algorithm vignette, grade 4

Mrs. Pottebaum recognized that daily practice of the subtraction algorithm was not helping Jesse make sense of or even learn the standard algorithm for subtraction. She decided that she must take the time to help Jesse make sense of some subtraction procedure, one that would let him compute accurately, even if it was not the standard subtraction algorithm.

Background. Jesse was staffed into a resource room for learning-disabled children during third grade. He needed constant monitoring and encouragement. The lengths of his assignments had to be modified. He was reluctant to use any form of recording, including drawing pictures and writing numbers. He had coordination problems, and his writing and drawings tended to be awkward and much too large for the space available on regular worksheets. He was willing to explain his thinking aloud, both to Mrs. Pottebaum and to the fourth-grade class, but he resisted writing anything. When he worked with a group of students, he benefited by having others do any recording. He was willing to participate but not to write.

Accommodations. Difficulty in writing seemed to be his main obstacle, so until Jesse gained more confidence, Mrs. Pottebaum tried to remove that barrier as much as possible. Despite a few minutes of daily review of basic skills, Jesse continued to have difficulty with the subtraction algorithm, particularly for problems involving renaming. He could not make sense of renaming and how one records it in the standard subtraction algorithm. Recognizing that Jesse would perform much better if he understood, Mrs. Pottebaum had the foresight to allow him to invent his own way to solve subtraction problems.

Mrs. Pottebaum organized her day so that she could find a few extra minutes to spend with Jesse each day. Instead of telling him how to solve subtraction problems, she listened to how he correctly solved easier subtraction problems. She learned that he did understand and could solve subtraction problems without renaming. He used a front-end approach, working with the hundreds or tens first. He also understood that he could "count up" or "add up" to subtract. To solve 423 – 195, Jesse subtracted 95 from 123 by finding how many more he would have to add to 95 to get 123. Mrs. Pottebaum capitalized on these understandings to help Jesse learn to deal with renaming, but she did so in a way that made sense to him. Instead of insisting that he first rename 423 as 3 hundreds, 12 tens, and 3 ones, and then 3 hundreds, 11 tens, and 13 ones, she let him use what made sense to him, which was to think of 423 as 300 + 123.

Breakthrough. After many extra sessions to help him develop a procedure that made sense to him, he could correctly solve subtraction problems as follows (in his words):

$$\begin{array}{r} 423 \\ -\ 195 \\ \hline \end{array}$$ "400 minus 100 is 300, but then I can't take 95 away from 23. So [*he started over*] I will take 100 away from 300 to get 200. I still have 123. Now I can take 95 away from that. From 95 to 100 is 5 and 23 plus 5 is 28. I had 200 left, so the answer is 228."

Summary. Jesse still did not want to write. He reluctantly drew pictures and wrote numbers to solve problems. Given larger pieces of paper, such as newsprint, he could write number sentences and even word problems that made sense. He could not complete nearly as many problems as other students in the same time.

The extra time that Mrs. Pottebaum gave him to make sense of subtraction conceptually paid huge dividends. He did not solve subtraction problems like everyone else, but he could solve them, explain his procedure, and understand how he did it. After having scored significantly below the level of an average third-grade student in computation in January of grade 3, Jesse scored 88 percent on a fourth-grade computational skills test one year later.

Distributed instruction with the freedom for Jesse to explore his own algorithm for subtraction problems let him make significant progress in computational achievement. He first made sense of a thinking strategy, which enabled him to make sense of a procedure for subtracting. Then he could solve the subtraction computation problems that were expected of the rest of the class. His procedure was different. He still did not want to record his process. But he was successful, and he knew it. He became much more confident.

Stage two: Recording a symbolic procedure

Daily previews before learning how to record a written skill should enable students to solve and explain computation problems specifically by using the thinking needed for the written skill. Then a few lessons will let the students learn how to record, with symbols, the actions and thinking that they now understand. With these understandings, helping students make these connections with symbols usually takes only two or three lessons. These lessons correspond to typical textbook lessons. In fact, teachers can use textbook lessons to help confirm students' abilities to solve computation problems symbolically. The students should also have to explain their thinking with the written skill throughout these lessons.

To help students make the connection between the actions they learned during the previews and writing the symbols in the algorithm, ask the meaning of each digit written in the algorithm. For example, when subtracting 423 – 195, ask how changing the 4 hundreds and 2 tens to 3 hundreds and 12 tens relates to the actions they used to solve this same problem with base-ten blocks. Continue with other digits in the written algorithm. Students should be able to explain each digit in the context of actions on the manipulatives. The purpose of stage two is for students to make sense of symbolically recording the concrete step-by-step actions they learned during the previews. Connecting the symbolic procedure to the thinking used to solve problems with concrete step-by-step actions enables the students to develop a deep understanding of symbolic procedures. Figure 5.9 shows sample activities to help students make these connections.

Stage three: Practice and connections for symbolic skills

After students have learned to record the written skill and can explain their thinking, brief reviews of the same skills should continue each day for two or three weeks, with explanations (and sometimes even concrete illustrations) for at least one problem. Attaining reasonable fluency with the skill usually requires no more than this for most students. Structured use of concrete materials can reinforce making sense of the skill for those students who still need those experiences.

You can use these previews and reviews to supplement your regular mathematics curriculum immediately before and after a unit where students must learn a skill.

After students have learned the written skill, they need continued experiences with the operation. Teachers must give them a variety of problems with different contexts, with different problem structures, and with applications to different situations. In particular, students must continue to develop connections to their mental computation and estimation skills. Although they now have a written skill to use for solving these problems, they need encouragement to use their reasoning strategies to solve the problems mentally as well. At this point the students should continue to develop flexibility in their thinking and strategic selection of appropriate solution strategies.

Summary of distributed instruction

Conceptual activities, word problems, and tasks distributed throughout the school year give students opportunities to learn multiple representations and multiple strategies to solve problems, which enable them to develop the number sense and flexibility they need for computational proficiency. Teachers should use these activities almost daily throughout the school year.

Even these number sense activities are not sufficient for many students to develop proficiency with skills. Some students also need focused instruction on the development of

Connecting a model to a standard algorithm might look like this:

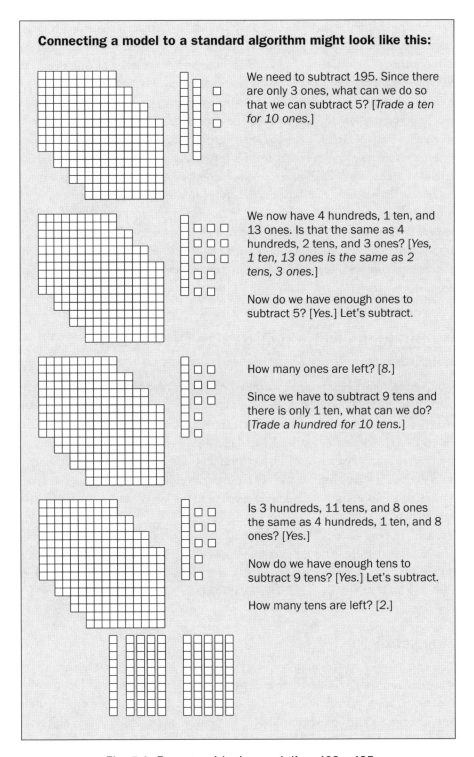

We need to subtract 195. Since there are only 3 ones, what can we do so that we can subtract 5? [*Trade a ten for 10 ones.*]

We now have 4 hundreds, 1 ten, and 13 ones. Is that the same as 4 hundreds, 2 tens, and 3 ones? [*Yes, 1 ten, 13 ones is the same as 2 tens, 3 ones.*]

Now do we have enough ones to subtract 5? [*Yes.*] Let's subtract.

How many ones are left? [*8.*]

Since we have to subtract 9 tens and there is only 1 ten, what can we do? [*Trade a hundred for 10 tens.*]

Is 3 hundreds, 11 tens, and 8 ones the same as 4 hundreds, 1 ten, and 8 ones? [*Yes.*]

Now do we have enough tens to subtract 9 tens? [*Yes.*] Let's subtract.

How many tens are left? [*2.*]

Fig. 5.9. Base-ten blocks modeling 423 – 195

proficiency of procedural skills beginning a few weeks before that skill is formally taught in the curriculum. These experiences include (1) previews to develop appropriate thinking for written algorithms; (2) lessons to connect that thinking to a symbolic record, that is, the written symbolic procedure; and (3) reviews to practice, explain, and connect the standard algorithm to other mental computation and estimation strategies. These brief conceptual previews (before a unit of instruction) and reviews of skills (after a unit of instruction), with structured activities to promote appropriate thinking, extend the time that students have to make sense of procedural skills, including standard algorithms, by several weeks. This extended time is crucial for students with learning problems. And the brief two- or three-minute activities do not prevent teachers from continuing their normal mathematics curriculum during this time. After this focus on the specific written skill, continued experiences enable the students to further develop flexibility in their thinking and strategic selection of solution strategies. Teachers should encourage students to continue to develop a variety of mental computational and estimation strategies to make sense of computational situations.

Using Big Ideas to Help Organize Distributed Instruction

Simply too many topics exist to use distributed experiences for every topic in the mathematics curriculum. Schools must choose two to three key topics or focal points that are essential number and operation topics for each grade level. Distributed instruction can become organized experiences for this limited number of topics. Understanding each key topic involves much more than learning procedures or skills. For students to understand a key topic, they must learn multiple ways to represent the idea, learn multiple strategies to solve problems, and become fluent in using these various tools so that their thinking is flexible and efficient.

Teachers can pose word problems related to a big idea during the distributed instruction. For example, addition and subtraction might be a big idea in the primary grades. Multiplication and division might be a big idea in the middle grades. Rational numbers might be a big idea in the upper grades. Proportional reasoning might be a big idea in middle school.

Number sense

Addition and subtraction are complementary key topics; in fact, the two can be integrated during the development of number sense. Subtraction problems should include a variety of take-away, missing-addend, and comparison situations. The problem structures should vary. Several contexts and applications should be present. See the following problem structures for addition and subtraction.

Part–Whole (If you know both parts, you can add to find the whole. If you know the whole and one part, you can subtract to find the other part.)

Action: join or separate

- JaRon had $4. He got $2 more for his birthday. How much money does he have now?
- JaRon had $4. He got some more money for his birthday. He now has $6. How much did he get for his birthday?
- JaRon had some money. He got $2 more for his birthday. He now has $6. How much money did he have before his birthday?
- Maria had 5 balloons. Two of them popped. How many balloons did she still have?
- Maria had 5 balloons. Some of them popped. She had 3 balloons left. How many balloons popped?
- Maria had some balloons. Two of them popped. She had 3 balloons left. How many balloons did she have at first?

No action

- James had 5 red pencils and 2 yellow pencils. How many pencils did he have?
- James had 7 red and yellow pencils. Two of them were yellow. How many were red?
- James had 7 red and yellow pencils. Five of them were red. How many were yellow?

Comparison (For the large set, if you know the part that matches the small set and the extra part, you can add to find the whole. If you know the whole and either the part that matches the small set or the extra part, you can subtract to find the other part.)

- Jo read 7 books. Ron read 5 books. How many more books did Jo read?
- Jo read 7 books. That was 2 more books than Ron read. How many books did Ron read?
- Ron read 5 books. That was 2 fewer books than Jo read. How many books did Jo read?

The numbers should vary to encourage different thinking strategies. For example, some students are likely to count back to solve "Jacob has $85. That is $2 more than he needs to buy new basketball shoes. How much do the shoes cost?" Some students are likely to count up to solve "Katie wants to send 37 valentines. She has already mailed some. She needs to mail 35 more. How many did she already mail?" Some students are likely to use known addition facts (20 + 20) to solve "Juan had to read 41 pages for an assignment. He read 20 pages. How many more pages did he have to read?"

Teachers should also use models or representations for subtraction repeatedly, including a variety of models and structured activities such as the ten frame, hidden parts, and the open number line. They give students new ways to represent subtraction situations and often encourage new thinking strategies. For example, the open number line in figure 5.10 provokes students to think about new ways to count up or add to find the difference between 51 and 28. Having only a few numbers marked on the number line encourages students to mark other numbers that will help them solve the problem. The figure includes samples of student thinking. In each case, the numbers in boldface are the numbers that these students might insert on the number line to help solve the problem.

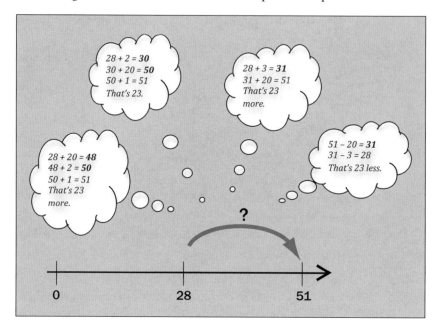

Fig. 5.10. Some thinking strategies that the empty number line provokes

Using several models to represent these problems and tasks and encouraging the students to solve the problems with multiple strategies will help them develop number sense, flexibility in thinking, and a variety of mental mathematics strategies.

Conceptual preview for basic facts

Some students will have developed efficient thinking strategies for solving basic subtraction facts. Others will still be counting inefficiently. To become fluent with the harder subtraction facts, these students must learn new thinking strategies to use efficiently with the harder facts.

For example, teachers can ask students to, with the open number line, "use ten" as a "stepping stone" to get from 8 to 13: "Two more is 10, then 3 more is 13. That's 5 more."

Creating situations in which some students explain and illustrate that thinking each day for about two or three weeks gives other students time to make sense of the new thinking and begin to use it with the harder subtraction facts. At the same time, they must extend this same thinking to situations with larger numbers. For example, "John has $48 saved to buy a new video game. It costs $53. How many more dollars does he need to save?" (Two more is 50 and 3 more is 53. That's 5 more.) Similar previews can help students make sense of counting up to subtract, counting back to subtract, using tens for mental computation (see fig. 5.3), and so on.

Practice and strategic choice for basic facts

After learning each of these mental computational thinking skills, students need to practice and explain these procedures. Two or three problems a day for two or three weeks, with explanations of at least one problem, will help students with learning problems integrate the new reasoning into their repertoire of skills. Teachers should encourage students to use them flexibly, depending on the numbers and the situation.

Concrete previews for symbolic skills

Before specifically teaching a written subtraction algorithm, teachers can briefly preview the thinking that will help the students make sense of the written skill. A standard subtraction algorithm involves renaming and then subtracting for each successive place value. To promote that same thinking, these previews can simply involve activities such as the following: Use base-ten blocks to show a number. Ask the students to explain how they could take away part of this number, but select numbers that involve renaming. During the discussion, lead them to rename a ten as 10 ones, if necessary, and then take away the ones; then rename 100 as 10 tens, if necessary, and then take away the tens; and so on. A similar problem each day for two or three weeks will enable most students to make sense of the thinking that they need for this standard algorithm. Symbols are not needed at this stage and they may interfere.

Recording symbolic procedures

Now that the students understand the step-by-step actions that they can use with base-ten blocks to solve subtraction computation examples and can offer coherent explanations, help them learn to record these actions symbolically. After renaming a ten as 10 ones, if necessary, record that renaming symbolically. Ask whether the renamed number is still the same value. Is 5 tens and 2 ones the same quantity as 4 tens and 12 ones? Then subtract the ones and record the difference in the ones place. Repeat this procedure for each place value. Each symbol written corresponds to actions on the base-ten blocks. Teachers should expect the students to explain that connection. To check, ask how each digit connects to the actions on the base-ten blocks.

Practice and connections for symbolic skills

After students learn the standard algorithm, have them practice that skill by using it to solve one or two problems each day for two or three weeks. The students should explain the connection between the actions and the symbols for at least one problem. Teachers can use concrete materials to illustrate the actions for those students who still need those experiences.

After this special attention to skills, teachers can pose problems and tasks daily for the rest of the school year. They need to present a variety of subtraction situations. During this time the students need encouragement to develop flexibility in their thinking and strategic selection of solution strategies. For example, teachers should encourage the students to learn to count up to make change with money. The empty number line might help provoke this thinking but can include target numbers, such as 25, 50, and 75 (fig. 5.11). Teachers should also challenge students to learn several ways to compute mentally and to estimate.

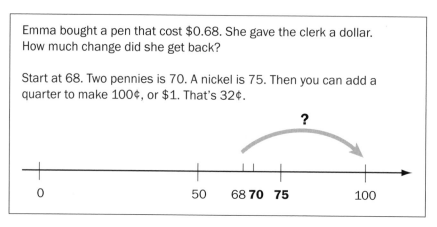

Emma bought a pen that cost $0.68. She gave the clerk a dollar. How much change did she get back?

Start at 68. Two pennies is 70. A nickel is 75. Then you can add a quarter to make 100¢, or $1. That's 32¢.

Fig. 5.11. Using the open number line to help children make change

Summary

Computational fluency requires both number sense and proficiency with skills. Students must develop a deep understanding of numbers and operations to achieve this goal. One way to give students the time they need to develop these understandings is to use distributed instruction with a focus on developing computational fluency. Doing so is particularly important for students with learning problems in mathematics. They need time to develop number sense. They need time to learn the thinking that will let them make sense of mental and symbolic procedures. The organization for planning curriculum that this chapter outlines gives these students opportunities to develop that number sense and flexibility and to develop proficiency with procedural skills, that is, to develop computational fluency and the confidence that grows with that accomplishment. Using distributed instruction and a focused curriculum to help all students develop computational fluency offers opportunities for students with learning problems in mathematics.

(Classroom vignettes by Angie Shindelar, Des Moines Community Schools, Greenfield, Iowa; Annette Louk, Prairie Lakes AEA 8, Pocahontas, Iowa; Melissa Cotton, Bettendorf Community Schools, Bettendorf, Iowa; and Pam Pottebaum, retired from Sioux City Community Schools, Sioux City, Iowa.)

Questions for Discussion

1. Students must develop an understanding of how computational procedures work. Far too often teachers show students, commonly those with learning needs, how to do a particular procedure without giving students the time to develop the pre-requisites for a procedural skill or understand how or why it works. How would you propose to develop computational fluency?

2. Suggest something you have done or currently do in your own classroom that supports the development of number sense. Why was it successful? How do you know? What have you read and discussed here that could add to a plan to ensure that all students have a sense of number?

3. How can we help students with inefficient counting strategies become proficient with basic multiplication facts? Discuss this chapter's treatment of this topic. How will it affect your own teaching of the basic facts?

4. Had you ever seen an open number line before? How might you use this important representational tool more effectively?

5. Proportional reasoning is a linchpin for the study of higher-level mathematics, particularly algebra. Read and discuss the vignette for grade 8 in this chapter. How did the teacher accommodate Tori's needs? What other interventions might the teacher have used?

References

Baxter, Juliet A., John Woodward, and Deborah Olson. "Effects of Reform-Based Mathematics Instruction on Low Achievers in Five Third-Grade Classrooms." *Elementary School Journal* 101 (May 2001): 529–47.

Baxter, Juliet A., John Woodward, Jill Voorhies, and Jennifer Wong. "We Talk About It, but Do They Get It?" *Learning Disabilities Research and Practice* 17 (August 2002): 173–85.

Bottge, Brian A., Enrique Rueda, Perry T. LaRoque, Ronald C. Serlin, and Jungmin Kwon. "Integrating Reform-Oriented Math Instruction in Special Education Settings." *Learning Disabilities Research and Practice* 22 (May 2007): 96–109.

Carnine, Doug, Eric D. Jones, and Robert Dixon. "Mathematics: Educational Tools for Diverse Learners." *School Psychology Review* 23, no. 3 (1994): 406–27.

Cepeda, Nicholas J., Harold Pashler, Edward Vul, John T. Wixted, and Doug Rohrer. "Distributed Practice in Verbal Recall Tasks: A Review and Quantitative Synthesis." *Psychological Bulletin* 132 (May 2006): 354–80.

Council of Chief State School Officers (CCSSO). *Common Core State Standards for Mathematics.* Washington, D.C.: CCSSO, 2010.

Fennell, Francis (Skip). *Number Sense Now! Reaching the NCTM Standards.* Reston, Va.: National Council of Teachers of Mathematics, 1993.

Fleischner, Jeanette E., Katherine Garnett, and Margaret Jo Shepherd. "Proficiency in Arithmetic Basic Facts Computation of Learning Disabled and Nondisabled Children." *Focus on Learning Problems in Mathematics* 4 (Spring 1982): 47–56.

Geary, David C., Carmen O. Hamson, and Mary K. Hoard. "Numerical and Arithmetical Cognition: A Longitudinal Study of Process and Concept Deficits in Children with Learning Disability." *Journal of Experimental Child Psychology* 77 (November 2000): 236–63.

Geary, David C., and Mary K. Hoard. "Numerical and Arithmetical Deficits in Learning-Disabled Children: Relation to Dyscalculia and Dyslexia." *Aphasiology* 15 (July 2001): 635–47.

Geary, David C., Mary K. Hoard, Jennifer Byrd-Craven, and M. Catherine DeSoto. "Strategy Use in Simple and Complex Addition: Contributions of Working Memory and Counting Knowledge for Children with Mathematical Disability." *Journal of Experimental Child Psychology* 88 (June 2004): 121–51.

Geary, David C., Mary K. Hoard, Lara Nugent, and Jennifer Byrd-Craven. "Strategy Use, Long-Term Memory, and Working Memory Capacity." In *Why Is Math So Hard for Some Children? The Nature and Origins of Mathematical Learning Difficulties and Disabilities*, edited by Daniel B. Berch and Michèle M. M. Mazzocco, pp. 83–105. Baltimore: Brookes Publishing, 2007.

Goldman, Susan R., James W. Pellegrino, and Davis L. Mertz. "Extended Practice of Fast Addition Facts: Strategy Change in Learning-Disabled Students." *Cognition and Instruction* 5 (September 1988): 223–65.

Jordan, Nancy C., and Teresa Oettinger Montani. "Cognitive Arithmetic and Problem Solving: A Comparison of Children with Specific and General Mathematics Difficulties." *Journal of Learning Disabilities* 30 (November–December 1997): 624–34.

Kroesbergen, Evelyn H., and Johannes E. H. van Luit. "Teaching Multiplication to Low Math Performers: Guided versus Structured Instruction." *Instructional Science* 30 (September 2002): 361–78.

Kroesbergen, Evelyn H., Johannes E. H. van Luit, and Cora J. M. Maas. "Effectiveness of Explicit and Constructivist Mathematics Instruction for Low-Achieving Students in the Netherlands." *Elementary School Journal* 104 (January 2004): 233–51.

Mazzocco, Michèle M. M. "Defining and Differentiating Mathematical Learning Disabilities and Difficulties." In *Why Is Math So Hard for Some Children? The Nature and Origins of Mathematical Learning Difficulties and Disabilities*, edited by Daniel B. Berch and Michèle M. M. Mazzocco, pp. 29–48. Baltimore: Brookes Publishing, 2007.

National Council of Teachers of Mathematics (NCTM). *Principles and Standards for School Mathematics.* Reston, Va.: NCTM, 2000.

———. *Curriculum Focal Points for Prekindergarten through Grade 8 Mathematics: A Quest for Coherence.* Reston, Va.: NCTM, 2006.

Pesk, Dolores D., and David Kirshner. "Interference of Instrumental Instruction in Subsequent Relational Learning." *Journal for Research in Mathematics Education* 31 (November 2000): 524–40.

Resnick, Lauren B., V. L. Bill, Sharon B. Lesgold, and M. N. Leer. "Thinking in Arithmetic Class." In *Teaching Advanced Skills to At-Risk Students: Views from Research and Practice*, edited by Barbara Means, Carol Chelemer, and Michael S. Knapp, pp. 27–53. San Francisco: Jossey-Bass, 1991.

Ritchhart, Ron, and David N. Perkins. "Life in the Mindful Classroom: Nurturing the Disposition of Mindfulness." *Journal of Social Issues* 56 (Spring 2000): 27–47.

Rohrer, Doug, and Kelli Taylor. "The Effects of Overlearning and Distributed Practice on the Retention of Mathematics Knowledge." *Applied Cognitive Psychology* 20 (December 2006): 1209–24.

Russell, Robert L., and Herbert P. Ginsburg. "Cognitive Analysis of Children's Mathematics Difficulties." *Cognition and Instruction* 1 (March 1984): 217–44.

Russell, Susan Jo. "Developing Computational Fluency with Whole Numbers." *Teaching Children Mathematics* 7 (November 2000): 154–58.

Schoenfeld, Alan H. "Making Mathematics Work for All Children: Issues of Standards, Testing, and Equity." *Educational Researcher* 31 (January–February 2002): 13–25.

Vaughn, Sharon, Janette Klingner, and Marie Hughes. "Sustainability of Research-Based Practices." *Exceptional Children* 66 (Winter 2000): 163–71.

Verschaffel, Lieven, Joke Torbeyns, Bert De Smedt, Koen Luwel, and Wim Van Doreen. "Strategy Flexibility in Children with Low Achievement in Mathematics." *Educational and Child Psychology* 24, no. 2 (2007): 16–27.

Woodward, John, and Juliet Baxter. "The Effects of an Innovative Approach to Mathematics on Academically Low-Achieving Students in Inclusive Settings." *Exceptional Children* 63 (Spring 1997): 373–88.

CHAPTER 6

Algebra

John K. Lannin
and Delinda van Garderen

This chapter focuses on the teaching and learning of algebraic concepts with K–8 students with special needs, including students with learning disabilities. We discuss the nature and purpose of algebraic thinking by considering the importance of learning this topic for children with special needs. The chapter also offers essential connections to the *Principles and Standards for School Mathematics* (National Council of Teachers of Mathematics [NCTM] 2000) and *Curriculum Focal Points for Prekindergarten through Grade 8 Mathematics: A Quest for Coherence* (NCTM 2006) for algebra, shares algebraic difficulties that students encounter, and suggests instructional techniques and modifications to meet the needs of students with disabilities. The suggested techniques and modifications also address the elements of algebra as defined in the operations and algebraic thinking, expressions and equations, and functions domains of the *Common Core State Standards* (Council of Chief State School Officers [CCSSO] 2010).

Research has paid much attention to algebra as a gatekeeper course (e.g., National Mathematics Advisory Panel 2008), drawing particular notice because of the focus on the development of mathematical reasoning before student entry into a middle school, junior high school, or high school course. Though people often view algebra as a middle school–junior high or high school course, in this chapter we focus on algebraic thinking rather than on algebra as a particular course. Algebraic thinking supports and often extends the typical view of algebra that many adults experienced as students. Algebraic thinking is closely connected to student learning in number and operations, furnishing support that can deepen student understanding of this content strand.

A large percentage of students have difficulty with the fundamental ideas related to algebra and algebraic thinking (Booth 1984; Küchemann 1981; Matz 1980). This chapter emphasizes informing about the nature of algebra and algebraic thinking because algebra has typically been viewed narrowly in the K–12 curriculum. To give a broader perspective on algebra, we focus on three aspects of algebra that Kaput (2008) delineated:

1. Building generalizations from arithmetic and quantitative reasoning
2. Generalizing patterns toward the idea of function, including determining whether two expressions are equivalent, where functions take on particular values, and whether they satisfy various constraints (building and solving equations)

3. Modeling various situations (creating generalizations that characterize the relationship between two quantities)

Because of the importance and connectedness of algebra and arithmetic, previous works have suggested that algebraic concepts develop across the K–12 curriculum rather than as a course detached from curriculum (NCTM 2000; Schoenfeld 1995). Because many students have difficulty with various aspects of algebraic thinking, this chapter emphasizes helping all students, including those with special needs, develop a deeper understanding of algebraic content.

Rationale for Focusing on Algebraic Thinking

Algebraic thinking supports developing student understanding of number and operation by fostering important ideas that underlie the basic facts for addition, subtraction, multiplication, and division. For example, in grades K–8 students must learn that addition and multiplication are commutative for both the whole numbers and the rational numbers and that subtraction and division are not commutative for these two sets of numbers. Students also learn and apply the distributive property for multiplication over addition (e.g., $3 \times 14 = 3 \times 10 + 3 \times 4$), building on the knowledge that they will apply in formal algebra [e.g., $3(x + 4) = 3x + 12$].

Students also learn to develop algebraic models for various situations, supporting their use of technological tools such as spreadsheets and computer programming. Developing an understanding of the meaning of variables in various situations is crucial for students and can encourage the use of quantitative reasoning that can in turn encourage a deeper understanding of number and operation.

Expectations and Needs Related to Algebra

Principles and Standards presents four overarching algebra standards for all K–12 students (NCTM 2000, p. 36):

1. Understand patterns, relations, and functions
2. Represent and analyze mathematical situations and structures by using algebraic symbols
3. Use mathematical models to represent and understand quantitative relationships
4. Analyze change in various contexts

Related to these four overarching standards for algebra, *Curriculum Focal Points* includes specific recommendations at each grade level related to deepening student knowledge of algebraic thinking (NCTM 2006, pp. 13–20). In the following we summarize the important focal points related to algebraic thinking.

Grade 1

Students use properties of addition {commutativity [i.e., that $a + b = b + a$] and associativity [i.e., that $(a + b) + c = a + (b + c)$]} to add whole numbers, and they create and use increasingly sophisticated strategies based on these properties (e.g., "making tens") to solve addition and subtraction problems involving basic facts. By comparing a variety of solution strategies, children relate addition and subtraction as inverse operations.

Through identifying, describing, and applying number patterns and properties in developing strategies for basic facts, children learn about other properties of numbers and operations, such as odd and even (e.g., "Even numbers of objects can be paired, with none left over") and zero as the identity element for addition.

Grade 2

Children develop, discuss, and use efficient, accurate, and generalizable methods to add and subtract multidigit whole numbers. They develop fluency with efficient procedures, including standard algorithms, for adding and subtracting whole numbers; understand why the procedures work (on the basis of place value and properties of operations); and use them to solve problems.

Grade 3

Students use properties of addition and multiplication {e.g., commutativity, associativity, and the distributive property [i.e., that $a(b + c) = ab + ac$]} to multiply whole numbers and apply increasingly sophisticated strategies based on these properties to solve multiplication and division problems involving basic facts. By comparing a variety of solution strategies, students relate multiplication and division as inverse operations.

Understanding properties of multiplication and the relationship between multiplication and division is a part of algebra readiness that develops at grade 3. The creation and analysis of patterns and relationships involving multiplication and division should occur at this grade level. Students build a foundation for later understanding of functional relationships by describing relationships in context with such statements as "the number of legs is four times the number of chairs."

Grade 4

Students apply their understanding of models for multiplication (i.e., equal-sized groups, arrays, area models, equal intervals on the number line), place value, and properties of operations (in particular, the distributive property) as they develop, discuss, and use efficient, accurate, and generalizable methods to multiply multidigit whole numbers. They develop fluency with efficient procedures, including the standard algorithm, for multiplying whole numbers; understand why the procedures work (on the basis of place value and properties of operations); and use them to solve problems.

Students continue identifying, describing, and extending numeric patterns involving all operations and nonnumeric growing or repeating patterns. Through these experiences, they develop an understanding of using a rule to describe a sequence of numbers or objects.

Grade 5

Students apply their understanding of models for division, place value, properties, and the relationship of division to multiplication as they develop, discuss, and use efficient, accurate, and generalizable procedures to find quotients involving multidigit dividends. They develop fluency with efficient procedures, including the standard algorithm, for dividing whole numbers; understand why the procedures work (on the basis of place value and properties of operations); and use them to solve problems. They consider the context in which a problem is situated to select the most useful form of the quotient for the solution, and they interpret it appropriately.

Students use patterns, models, and relationships as contexts for writing and solving simple equations and inequalities. They create graphs of simple equations. They explore prime and composite numbers and discover concepts related to the addition and subtraction of fractions as they use factors and multiples, including applications of common factors and common multiples. They develop an understanding of the order of operations and use it for all operations.

Grade 6

Students write mathematical expressions and equations that correspond to given situations, they evaluate expressions, and they use expressions and formulas to solve problems. They understand that variables represent numbers whose exact values are not yet specified, and they use variables appropriately. Students understand that expressions in different forms can be equivalent, and they can rewrite an expression to represent a quantity in a different way (e.g., to make it more compact or to feature different information). Students know that the solutions of an equation are the values of the variables that make the equation true. They solve simple one-step equations by using number sense, properties of operations, and the idea of maintaining equality on both sides of an equation. They construct and analyze tables (e.g., to show quantities that are in equivalent ratios), and they use equations to describe simple relationships (e.g., $3x = y$) shown in a table.

Students use the commutative, associative, and distributive properties to show that two expressions are equivalent. They also illustrate properties of operations by showing that two expressions are equivalent in a given context [e.g., determining the area in two different ways for a rectangle whose dimensions are $(x + 3)$ by 5]. Sequences, including those that arise in the context of finding possible rules for patterns of figures or stacks of objects, give students opportunities to develop formulas.

Grade 7

Students extend their work with ratios to develop an understanding of proportionality that they apply to solve single-step and multistep problems in many contexts. They use ratio and proportionality to solve a wide variety of percent problems, including problems involving discounts, interest, taxes, tips, and percent increase or decrease. They also solve problems about similar objects (including figures) by using scale factors that relate corresponding lengths of the objects or by using the fact that relationships of lengths within an object are preserved in similar objects. Students graph proportional relationships and identify the unit rate as the slope of the related line. They distinguish proportional relationships ($y/x = k$, or $y = kx$) from other relationships, including inverse proportionality ($xy = k$, or $y = k/x$).

By decomposing two- and three-dimensional shapes into smaller, component shapes, students find surface areas and develop and justify formulas for the surface areas and volumes of prisms and cylinders. As students decompose prisms and cylinders by slicing them, they develop and understand formulas for their volumes (volume = area of base × height). They apply these formulas in problem solving to determine volumes of prisms and cylinders. Students see that the formula for the area of a circle is plausible by decomposing a circle into several wedges and rearranging them into a shape that approximates a parallelogram. They select appropriate two- and three-dimensional shapes to model real-world situations and solve a variety of problems (including multistep problems) involving surface areas, areas and circumferences of circles, and volumes of prisms and cylinders.

Students extend understandings of addition, subtraction, multiplication, and division, together with their properties, to all rational numbers, including negative integers. By applying properties of arithmetic and considering negative numbers in everyday contexts (e.g., situations of owing money or measuring elevations above and below sea level), students explain why the rules for adding, subtracting, multiplying, and dividing with negative numbers make sense. They use the arithmetic of rational numbers as they formulate and solve linear equations in one variable and use these equations to solve problems. Students make strategic choices of procedures to solve linear equations in one variable and implement them efficiently, understanding that when they use the properties of equality to express an equation in a new way, solutions that they obtain for the new equation also solve the original equation.

Grade 8

Students use linear functions, linear equations, and systems of linear equations to represent, analyze, and solve a variety of problems. They recognize a proportion ($y/x = k$, or $y = kx$) as a special case of a linear equation of the form $y = mx + b$, understanding that the constant of proportionality (k) is the slope and the resulting graph is a line through the origin. Students understand that the slope (m) of a line is a constant rate of change, so if the input, or x-coordinate, changes by a specific amount, the output, or y-coordinate, changes by the amount ma. Students translate among verbal, tabular, graphical, and algebraic representations of functions (recognizing that tabular and graphical representations are usually only partial representations), and they describe how such aspects of a function as slope and y-intercept appear in different representations.

Students solve systems of two linear equations in two variables and relate the systems to pairs of lines that intersect, are parallel, or are the same line in the plane. Students use linear equations, systems of linear equations, linear functions, and their understanding of the slope of a line to analyze situations and solve problems.

Students encounter some nonlinear functions (such as the inverse proportions that they studied in grade 7 as well as basic quadratic and exponential functions) whose rates of change contrast with the constant rate of change of linear functions. They view arithmetic sequences, including those arising from patterns or problems, as linear functions whose inputs are counting numbers. They apply ideas about linear functions to solve problems involving rates such as motion at a constant speed.

The focal points serve as a developmental trajectory for algebraic thinking, guiding the deepening of student understanding of important ideas related to number and operation over the course of kindergarten through grade 8. Such background is intended to develop a firm foundation of background knowledge so that all students can have access to higher levels of mathematics in secondary school and beyond.

Building Generalizations from Arithmetic and Quantitative Reasoning

As young children engage in situations that they can model using addition or subtraction, opportunities arise to assist all children connecting their initial understanding of number and operation to general ideas that exemplify algebraic thinking. For example, Carpenter, Franke, and Levi (2003) describe various situations that elementary-level students can engage in that connect to algebraic thinking. Consider the following numeric situations:

$$25 + 4 - 4 = ?$$
$$25 + 10 - 10 = ?$$
$$25 + 17 - 17 = ?$$

Students can generalize that "adding and subtracting the same thing does not change the original value." We could generalize this concept by using formal algebraic symbols as $x + y - y = x$, where x and y represent any real number. Of course, at this age, students cannot generalize this situation to the real numbers but can recognize that this generalization is true for all the whole numbers that they know. Students can use various representations (e.g., base-ten blocks) to demonstrate how, after starting with an initial quantity, adding and removing the same quantity of blocks does not change the initial quantity.

Teachers can use other, similar situations for children to examine the commutative and associative properties of addition and multiplication, properties of even and odd numbers, and the distributive property of multiplication over addition. In each situation, teachers should encourage students to examine examples of these properties, engage in conjecture related to generalizations, and explain why such a generalization would be true.

The challenge with elementary school students (and continuing up to students in college) involves assisting them in recognizing what constitutes a valid and invalid argument to support a generalization that involves an infinite domain (e.g., for all whole numbers). Students offer justifications that fall into three categories: no justification/appeal to authority, empirical justification, or generic examples (Carpenter, Franke, and Levi 2003; Simon and Blume 1996). Students using the first category appear to believe that generalizations require no explanation (e.g., it just is true that 3×5 is the same as 5×3) or believe a statement is true because a credible person says so. Students who justify often rely on a few examples to determine whether a general statement is true (empirical justification). For example, students may examine 3×5 and 5×3, 2×6 and 6×2, and 8×4 and 4×8 and conclude that one can multiply all whole numbers in any order. Such a justification is mathematically problematic because it relies on a few instances rather than a general relationship to demonstrate why a generalization is always true. To understand why this is problematic, consider the statement "all even numbers can be divided evenly by 4." An infinite number of even numbers are divisible by 4 (e.g., $4, 8, 12, 16, 20, \dots$). However, not all even numbers are divisible by 4 because numbers such as 6 and 10 are even, but when one divides them by 4 the result is not a whole number. Even though we can generate many examples of even numbers that are evenly divisible by 4, the statement is not true for all even numbers. Students at the elementary school level tend to rely on empirical justification because they can offer no other means for justifying their general statements. However, K–8 students can begin to examine general arguments that support the general relationships that they identify. Students could represent these instances by examining the partitioning of quantities into four equal-sized groups of discrete objects. They could examine that they can place twenty objects into four equal groups of five but that they cannot place six objects into four equal groups without further partitioning the object.

The final category of justification, generic examples, lends itself to acceptable justification of general statements if instruction emphasizes the general relationship and not the specific instances. Consider the statement that for any whole number, you can add in either order (e.g., $4 + 5 = 5 + 4$). How could students properly justify this statement? If they recognize that quantity is conserved (i.e., that changing the location of objects does not affect the number of objects) then they can state that combining four blocks with five blocks is the same as combining five blocks with four blocks. Which quantity is added to the other does not matter as long as the two quantities are combined. This explanation includes an example (i.e., that $4 + 5$ is the same as $5 + 4$); however, a general argument is given related to combining quantities: that rearranging objects does not affect the final result. The example is generic in that it suggests an overarching idea that one could apply to all examples of adding whole numbers.

In the elementary school classroom, other aspects of generalization arise as students begin to explore algorithms for addition, subtraction, multiplication, and division of whole numbers (and later on fractions and decimals). Introducing algorithms gives all students opportunities to examine efficient strategies for the operations. Typically educators have asked students to repeat algorithms even though the students may not understand why the algorithms do or do not generate correct results for many situations. However, a study of various algorithms and why they yield the correct results for all values in a given domain can build student algebraic understanding. Again, such reasoning involves a general understanding of number and number relationships.

Consider the subtraction algorithm for whole numbers that is taught in much of Latin America (often referred to as the Austrian method) for 63 – 28 (fig. 6.1). Similar to the traditional U.S. algorithm, we consider the result of subtracting 8 ones from 3 ones. Since this result will not generate a positive whole number, we add ten to the 3 ones, resulting in 13 ones. In addition, to maintain the same difference between 63 and 28, we add a ten to 28, resulting in 38. Now we can subtract the ones (13 – 8 = 5) and subtract the tens (60 – 30 = 30), finding a difference of 35. Such a procedure relies on an important mathematical idea: that the difference between two numbers is constant when one adds the same quantity to each value. Why is this true?

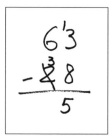

Fig. 6.1. Standard subtraction algorithm
used in Latin America

To grasp why one can apply this procedure to all whole numbers, understanding a general idea—one even more general than the one applied here—is helpful. Consider how the results of 123 – 75 and 125 – 77 compare. If you stated that these are the same, consider why that is. Comparing 123 – 75 and 125 – 77 on the open number line is one way to evaluate these results. One model for 123 – 75 involves examining how far 123 is from 75 on the number line. The distance between 123 and 75 is the same as that between 125 and 77; a shift of two units has occurred between the two values, preserving the distance between them. This generic example illustrates why adding (or subtracting) the same quantity from each value would always result in the same difference. We encourage using various models to help students understand why they can apply the procedures that they learn for addition, subtraction, multiplication, and division to all whole numbers (or rational numbers) that they use. Students can also engage in quantitative reasoning in many other situations (fig. 6.2). See Carpenter, Franke, and Levi (2003) for more examples of situations that students can examine in the elementary grades.

Teachers can use many different representations to assist students in representing and recognizing generality. However, one challenge of algebraic thinking is that no object exists that demonstrates a general case (Radford 2000). For example, in the previous open number line model, we compared the results for 123 – 75 and 125 – 77. However, no general model exists that can show why adding 2 (or any quantity) to both the minuend and the subtrahend will always generate the same result. However, our minds can recognize general relationships, and we can describe and justify these. Such relationships point to the importance of visualization in encouraging students to deepen their understanding of algebraic ideas.

Example 1:
- Which is more: 35 × 37 or 36 × 36?
- Which is more: 29 × 31 or 30 × 30?
- Which is more: 47 × 49 or 48 × 48?
- What is true in general in these situations? Why is this always true?

Example 2:
- Which is more for your bill, adding 7% tax first and taking a 20% discount or taking a 20% discount and adding 7% tax?
- Which is more for your bill, adding 5% tax first and taking a 30% discount or taking a 30% discount and adding 5% tax?
- Which is more for your bill, taking a 30% discount first and then taking a 20% discount or taking a 20% discount first, followed by a 30% discount?
- What is true in general in these situations? Why is this always true?

Fig. 6.2. Examples that involve examining quantitative general reasoning

Carraher and Schliemann (2007) note the importance of using various ways to examine and express generalizations: "Conventional algebraic notation is not the only vehicle for the expression of algebraic ideas and relations; tables, number sentences, graphs and specialized linguistic structures can also express algebraic ideas" (p. 624). Tables, graphs, the open number line, arrays, base-ten blocks, and number sentences all offer means to express and assist students in identifying and expressing generalizations. However, we must recognize the limitations of each representation. Many representations, such as tables, number sentences, and the open number line, generally allow for examining only a finite number of cases. For example, using tables can focus students on number relationships absent a connection to meaning or context (Swafford and Langrall 2000). These tools can assist students in developing generalizations but can also distract students or promote a disconnection from the meaning and explanation that we want students to present for their generalizations.

Furthermore, representations can offer various means for expressing the generalizations they identify. The developmental stages of expressing generalization tend to mirror the historical development of algebra, building from an initial rhetoric stage, to a syncopated stage, to a formal symbolic stage (Boyer and Merzback 1989; Harper 1987; Swafford and Langrall 2000). The initial rhetorical stage involves verbally describing a generalization. For example, stating that "an odd plus an odd is an even" is a common means for elementary-level students to describe their generalizations. In the upper elementary grades, as students begin to familiarize themselves with formal mathematical symbols, they often invent syncopated expressions, such as $3 \times \_\_ + 4$ to describe multiplying 3 by any number and adding 4. Teachers can then introduce students to formal symbols for algebra, such as $3y + 4$, to describe the same expression or $x + y = y + x$ to represent two expressions that are always the same. However, students have demonstrated considerable difficulty with developing meaning for algebraic symbols, and consistent reference to the meaning of these symbols as well as connecting verbal, syncopated, and formal symbols may better assist students in developing meaning for algebra notations (Lannin et al. 2008).

Generalizing Patterns toward the Idea of Function

"A function is a rule that assigns each element from a domain to a unique element in the co-domain" (Carraher and Schliemann 2007, p. 688). Working with functions involves examining various mathematical functions (e.g., cost [in dollars] = $4x$, where x is the number of boxes of cereal purchased) and considering how these functions increase or decrease as the variables increase or decrease. A deep understanding of functions also includes recognizing that two different-looking expressions can generate the same result for all values in a particular domain (often the real numbers, though at the elementary and middle school level, this will include the whole numbers and positive rational numbers).

Students encounter functions in their experiences outside school. They recognize that the grocery store charges them one specific amount for purchasing a particular number of items. They know that if they purchase several boxes of cereal at $4 per box, they can find the total cost by multiplying the number of boxes by 4. With some discussion, they can recognize that increasing the number of boxes purchased by 1 increases the total cost by 4. Though students need not learn the formal terminology related to functions and formal function notation [e.g., $f(x) = 4x$], teachers can and should introduce them to these ideas. Such tasks introduce students to rates of change, which students can examine through using various representations (e.g., input–output machines, function tables).

Another means of introducing students to functions occurs through examining two equivalent, but different-looking, functions. For example, students may identify the perimeter of a rectangle by using the rule $P = 2a + 2b$ (where a and b are the side lengths of the rectangle), whereas other students may represent the perimeter with $P = 2(a + b)$. Teachers can and should encourage students to consider the values for which the expressions $2a + 2b$ and $2(a + b)$ are equivalent and why they are equivalent for all values. Encouraging students to connect to the meaning of these expressions [e.g., recognizing that $2a + 2b$ is equivalent to $a + a + b + b$ and that $2(a + b)$ is equivalent to $(a + b) + (a + b)$] can help them recognize why these two expressions are equivalent. Students may want to verify that two expressions are equivalent by testing to see whether they yield the same values for a few instances. As discussed previously, teachers should encourage students to see the limitations of such arguments and the validity of general arguments such as the preceding connection to meaning.

A final and often-emphasized task in algebra involves examining when a function takes on particular values. Students can connect this concept to solving equations, much of which has been the traditional focus early in algebra courses. Consider the problem from Nathan and Koedinger (2000, p. 170): "When Ted got home from his waiter job, he multiplied his hourly wage by the 6 hours he worked that day. Then he added the \$66 he made in tips and found he earned \$81.90. How much did Ted make per hour?" When Nathan and Koedinger examined student performance on contextual situations compared to similar symbolic representations of these situations (e.g., solve for x: $6x + 66 = 81.90$), performance of students in a formal algebra course was higher for the contextual problems than for those represented symbolically. Given contextual situations, students used their own natural strategies that made sense to them, whereas they often confused the rules they learned in their formal algebra classes. Encouraging students to solve context-based problems, such as those that Nathan and Koedinger introduced, and encouraging students to solve these problems by using their own strategies, followed by connections to formal algebraic strategies (if appropriate for the grade level), appears to build on student knowledge gained in earlier grades.

Connecting various representations of functions can also deepen student understanding of the meaning and the use of functions. For example, students should examine the rate of change of functions so that they begin to recognize how graphical, symbolic, and tabular forms of linear functions demonstrate a constant rate of change and how they differ from nonlinear functions. They can also investigate how the symbolic, graphical, and tabular representations of equivalent expressions relate to one another, recognizing the limitations of the graphical and tabular representations for justifying the equivalence of expressions.

Modeling Situations

Many studies have examined the general mathematical rules and models that students can generate for different types of contexts. Students of various ability levels have demonstrated the ability to generate and apply strategies for these contexts. Creating generalizations for situations such as the cube sticker problem (fig. 6.3) can encourage students to (1) connect operations (e.g., addition and multiplication), (2) develop a view of variables as varying quantities, and (3) examine equivalent or nonequivalent expressions. For the cube sticker problem, some students may generate the rule $S = 4n + 2$, where S is the number of stickers and n is the number of cubes. Other students may develop the rule $S = 10 + 4(n - 2)$. The former rule describes the two stickers needed for each end of the rod and the four stickers needed per cube for the lateral faces of each cube. The latter rule is similar but calls for counting the five stickers for the two cubes on the ends of the rod before adding the number of stickers on the lateral faces of the remaining cubes. Students generate a wide variety of rules for such situations, allowing for multiple access points for such situations for all students (e.g., Kenney, Zawojewski, and Silver 1998).

Researchers (Healy and Hoyles 1999; Stacey 1989; Swafford and Langrall 2000) have identified various modeling strategies that students use (table 6.1). Other researchers (Stacey and MacGregor 2001; Swafford and Langrall 2000) have demonstrated that some

A company makes colored rods by joining cubes in a row and using a sticker machine to place "smiley" stickers on the rods. The machine places exactly one sticker on each exposed face of each cube. Every exposed face of each cube has to have a sticker, so this length-2 rod would need 10 stickers.

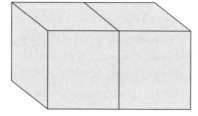

1. How many stickers would you need for rods of length 1–10? Explain how you determined these values.

2. How many stickers would you need for a rod of length 20? Of length 50? Of length 127? Explain how you determined these values.

3. Explain how you could find the number of stickers needed for a rod of any length. Write a rule that you could use to determine this.

Fig. 6.3. Cube sticker problem

students tend to focus only on numeric relationships (e.g., examining the increase in the number of stickers in a table for the cube sticker problem) without connecting this reasoning to the visual representation (e.g., considering how inserting a cube at the end of a rod will increase the number of stickers). As Healy and Hoyles noted, connecting numeric relationships to iconic/visual representations in situations can increase student success with developing appropriate generalizations.

Table 6.1
Framework of students' strategies for generalizing problem situations[a]

Strategy	Iconic	Numeric
Explicit	An explicit rule is constructed based on a visual representation of the situation by connecting to a counting technique (e.g., there are four stickers for each cube, so I took four times the length of the rod and then I added two stickers for the ends of the rod).	The student guesses an explicit rule based on a numeric pattern in the output values (e.g., since the number of stickers needed for a length-2 rod is 10, I can multiply the length of the rod by 5 to obtain the number of stickers).
Whole object (also "unitizing")	The student uses a portion as a unit to construct a larger unit by using multiples of the unit. The student adjusts for overcounting (or undercounting) due to the visual overlap that occurs when units are connected (e.g., a rod of length 10 has 42 stickers, so a rod of length 20 would have 42 × 2 − 2 because the stickers between the two length-10 sections would need to be removed).	The student uses a portion as a unit to construct a larger unit by using multiples of the unit. The student fails to adjust for any over- or undercounting, where applicable (e.g., a rod of length 10 has 42 stickers, so a rod of length 20 would have 42 × 2, or 84, stickers [incorrect]).
Chunking	A recursive rule is established based on a relationship established in the diagram, adding a unit onto known values of the desired attribute [e.g., for a rod of length 10 there are 42 stickers, so for a rod of length 5, I would take 42 + 5(4) because each cube adds four stickers].	The student builds on a recursive pattern by referring to a table of values, building a unit onto known values of the desired attribute [e.g., for a rod of length 10 there are 42 stickers, so for a rod of length 5, I would take 42 + 5(4) because the number of stickers increases by 4 each time].
Recursive	The student describes a relationship that occurs in the visual relation of the situation between consecutive values of the independent variable (e.g., each additional cube adds 5 stickers, and one sticker must be removed when the new cube is added to the rod, making a total of 4 stickers added for each cube).	The student notices a number pattern in the results for consecutive values of the independent variable (e.g., the number of stickers goes up 4 each time when the length of rod is increased by 1. The number of stickers goes 6, 10, 14, 18, etc.).

[a]Adapted from Healy and Hoyles (1999).

When reasoning recursively, a student could examine consecutive output values for the cube sticker problem (see fig. 6.3). By using a physical model and counting, the student could determine the number of stickers for a length-4 rod and a length-5 rod and notice that the number of stickers increases by 4 each time. At this point, the student moves from examining particular instances to considering a general relationship. After realizing that inserting a cube between two blocks in the rod adds four stickers, the student recognizes that this action could work for any length of rod and concludes that increasing the length of a rod by 1 always increases the number of stickers by 4. In this situation the student could begin by reasoning using a recursive strategy, examine how the numeric relationship of adding 4 related to the visual representation of the situation, and draw a general conclusion about the recursive relationship.

Students can also apply explicit reasoning for this situation. For example, a student could consider how to count the number of stickers on a length-10 rod. Noticing that the rod contains 10 stickers on the top, bottom, and two sides, and an additional 2 stickers on the ends of the rod, lets the student recognize that this strategy could work for all rods, leading to the development of the explicit rule $S = 4n + 2$, where S is the number of stickers and n is the length of the rod. Here the student began reasoning by recognizing an iconic relationship and applied this reasoning to the general case through using an explicit rule.

The student could also use recursive/explicit reasoning. Using the fact that the number of stickers increases by 4 each time, students can build on the knowledge that one cube contains six stickers. Each added cube would increase the number of stickers by 4, leading to developing the rule $S = 6 + 4(n - 1)$.

As mentioned previously, students need not use formal algebraic symbols to represent their rules. They can use verbal rules (e.g., four times the length of the rod plus two) or syncopated rules (e.g., $4 \times$ length of rod $+ 2$) that are equivalent to the written rules. Whenever students model situations mathematically, teachers must encourage them to define the variables that they use, ask them what values of the variables are appropriate for the situation, and require them to justify their generalizations.

Commonly Identified Student Difficulties and Misconceptions with Algebraic Symbols

Symbols play an important role in learning mathematics. People often view algebra as a language focused on using particular symbols. However, the mathematical ideas that the symbols express are the primary aspect of algebraic thinking. Here we focus on the development of meaning for algebraic symbols and discuss how we can help all students do so.

Many students, through their school experiences, develop a narrow perspective of what it means to do mathematics, particularly for algebra (Kieran 2007; Sfard and Linchevski 1994). For example, students often view the symbolic representations of variables (i.e., x, y, m) as "unknowns" rather than varying quantities. Thus, children tend to engage in "finding x" in various situations that may or may not involve using a variable as an unknown. Such a view of variables leads to difficulties when children come to equations such as $x + y = y + x$. Here students often find such tasks confusing. Algebraic symbols represent different quantities in different situations. Adding to the confusion, the use of algebraic symbols often shifts as students use them in problem situations. Consider determining a linear equation to represent the cost, y, of x marbles at \$2 each in addition to a fixed shipping charge of \$3. A student could start with the standard linear equations ($y = mx + b$; x and y are initially varying quantities, and m and b are parameters). Replacing m and b with constants shifts how they are used, from symbols to constants, resulting in the equation $y = 2x + 3$. If we wish to find how many marbles we can order for $31, y is suddenly a fixed value for this instance, and x becomes an unknown in the equation $31 = 2x + 3$. Such a shift in meaning and use of symbols requires explicit attention regarding symbol usage.

A diminished understanding of any symbol can lead to difficulties in developing meaning and mathematical sense making. Some symbols, such as those for the operations and the equality symbol (=), serve as the primary symbols for representing quantitative relationships in the elementary grades. In particular, how students view the equality symbol is a difficult cognitive obstacle (Booth 1984; Kieran 1981). Student difficulty with developing meaning for the equality symbol emerges as early as kindergarten and can continue throughout elementary school (and probably beyond). Consider $4 + 5 = \_\_ + 2$. Children believe primarily that the values that make the equation true are 9 (i.e., adding 4 and 5 and ignoring the 2) or 11 (i.e., combining 4, 5, and 2). These children see the equality symbol, which implies that they must write the result or answer for this equation. They have difficulty making sense of equations written in what they view as nonstandard ways (e.g., $9 = 4 + 5$, $6 = 6$, or $3 + 3 = 4 + 2$). This view appears to emerge early in children's thinking and can be quite resilient. Prolonged focus on the appropriate use and meaning of the equality symbol is necessary to help children progress to a broader meaning of this symbol. See Carpenter, Franke, and Levi (2003) for further discussion about the appropriate and inappropriate uses of the equality symbol in classrooms as well as a discussion of targeted instructional strategies that teachers can use to help students shift in their view of equality. Carefully constructed activities have demonstrated that children as young as eight and nine years can adopt appropriate meaning of the equals sign. Diminished understanding in middle school students of the meaning for the equals sign can lead to difficulties with solving algebraic equations (Knuth et al. 2006).

Küchemann (1981) reported that high school–aged students had difficulties representing word problems with equations. In his study, students struggled to correctly produce a cost equation when given two colors of pencils with different costs. Many students viewed the variables as representing labels for the different sets of pencils (i.e., r for red and b for blue) rather than as symbols that represented varying quantities. Küchemann (1981) and Booth (1984) identified a few other common misconceptions about mathematical symbols:

1. Thinking that the variables x and y cannot both have the same value
2. Viewing variables as labels (e.g., m = money rather than the amount of money)
3. Believing that 1 more than x is y
4. Viewing a as 1, b as 2, c as 3, and so on

For students to develop an appropriate meaning for variables, teachers should ask them to describe what the symbols mean and to refer to the domain of the variable. For example, in the previous equation $y = 2x + 3$, where x represents the number of marbles and y is the cost, x could be any natural number (i.e., 1, 2, 3, . . .). But for x here to be a rational number that is not a whole number (e.g., $^3/_4$) would not be reasonable because we would not use such a number for the number of marbles. Using zero for x is not reasonable since we would not charge an additional fee for the cost of ordering zero marbles. Explicit discussion about the use and meaning of variables can assist students by making transparent what the symbols that they use represent. A key aspect of algebraic thinking involves helping students build an understanding of variables as a varying quantity and the use of mathematical expressions and equations related to their underlying meanings.

Challenges that Students with Disabilities Experience

In mathematics many students with disabilities perform considerably lower than their typically developing peers, often demonstrating little progress from one year to the next (Cawley and Miller 1989; Cawley et al. 2001; Deshler et al. 2004; Wirt et al. 2004). Interestingly, to our knowledge no study has examined specific difficulties that students with disabilities may experience when doing algebra. However, the following common difficulties that students with disabilities experience when doing mathematics may also interfere with their ability to do algebra. These include (1) difficulty retaining and recalling basic facts; (2) difficulty constructing a representation of problems, including distinguishing relevant from irrelevant information; (3) computational deficits, including selecting appropriate operations for a given problem as well as executing the numerical calculation; (4) language deficits such as reading and understanding key terms; (5) weak abstract reasoning skills; (6) difficulties self-monitoring performance both for self-regulating what to do to solve a problem and for checking and self-correcting errors; and (7) lack of motivation, low self-esteem, or low self-efficacy for doing mathematics, often due to repeated academic failure (Gagnon and Maccini 2001; Ives and Hoy 2003; Maccini, McNaughton, and Ruhl 1999; Miles and Forcht 1995; Steele and Steele 2003).

Instructional Recommendations from the Field of Special Education

The special education literature offers limited instructional methods in algebra specifically for students with disabilities. Across the available studies, two main recommendations exist: (1) using explicit and/or direct instruction and (2) using peers to support learning.

Explicit and/or direct instruction

Explicit or direct instruction in the special education literature typically connotes one of two meanings. The first is using a systematic instructional routine, which involves explaining and demonstrating specific strategies (e.g., a step-by-step procedure or a strategy such as using representation) to solve a problem. Second, this approach can also involve how the teacher sequences problems to highlight a crucial skill or concept (National Mathematics

Advisory Panel 2008). Both ideas, however, are often combined.

Essential components of a systematic instructional routine include giving students an advance organizer, which describes what they will be learning for the day and why it is important; modeling or demonstrating a procedure with many examples; guided or structured practice in a whole-class or small-group format focused on using problems similar to those in the demonstration; independent practice with teacher support; corrective feedback such as giving cues and prompts reminding the students of the steps to complete the problem during all practice opportunities; and multiple opportunities for practice and review of the content (Allsopp 1999; Gagnon and Maccini 2001; Maccini and Ruhl 2000; Steele and Steele 2003; Witzel, Mercer, and Miller 2003).

The special education literature on algebra has promoted three main types of strategies to focus on in instruction. First is using step-by-step procedures for solving particular types of problems. Two recommended procedures include using a cognitive strategy called STAR (search, translate, answer, and review; Allsopp 1999; Gagnon and Maccini 2001; Maccini and Hughes 2000; Maccini and Ruhl 2000) and a graphic organizer (Ives 2007; Ives and Hoy 2003). STAR involves four steps that students apply when solving various algebra problems:

1. Search the word problem, where students read the problem and write down what is known and unknown.
2. Translate the words into an equation in a picture form by using some kind of manipulative.
3. Answer the problem.
4. Review the solution by rereading the problem and checking the reasonableness of the answer generated.

This strategy reflects the classic Pólya (1957) problem-solving heuristic, in which a learner should (1) first understand the problem, (2) devise a plan consisting of a series of moves for solving the problem, (3) carry out the plan, and (4) look back to check whether he or she in fact solved the problem. The graphic organizer consists of a rectangle divided into six sections that the students can use to solve linear equations as they work clockwise from cell to cell, starting in the top-left cell.

The second recommended strategy is to encourage students to use a representation to aid in understanding the problem and to offer a means for solving it. A recommended instructional method for promoting a hands-on experience that involves representations is the CRA (concrete to representational to abstract) sequence (Witzel 2005; Witzel, Mercer, and Miller 2003; Witzel, Riccomini, and Schneider 2008; Witzel, Smith, and Brownell 2001). This instructional sequence teaches students to solve a problem by using three representational stages: (1) the concrete stage, where students represent the problem with manipulatives such as sticks and cups; (2) the pictorial stage, where students translate the concrete representation into a pictorial representation; and (3) the abstract stage, where students translate the pictorial representation into abstract representation through using arabic numbers and operational symbols. Educators explicitly teach each stage of the sequence to students (i.e., teaching them how to represent the problems with the representation) along with instruction to help students understand how the sequence is interconnected.

A third recommended strategy is to teach students how to self-monitor and self-check what they are doing. Hutchinson (1993) taught students to use a set of questions to ask themselves while representing (e.g., "Have I read and understood each sentence? Have I got the whole picture, a representation, for this problem?") and solving (e.g., "Have I written an equation? Have I expanded the terms? What should I look for in a new problem to see if it is the same kind of problem?") algebra word problems. Such a metacognitive strategy is likely to assist all students in thinking about their own reasoning in various problem situations. The following questions are particularly important when encouraging algebraic thinking: "What does the variable mean?" "Is this statement true for all values that have been identified?" "How do I know that this statement is true for all values?"

Use of peers

Special education researchers have also promoted using peers (Allsopp 1997, 1999; Kortering, deBettencourt, and Braziel 2005), although they have not as extensively examined this approach as they have explicit or direct instruction. One peer-tutoring format suitable for heterogeneous classrooms is classwide peer tutoring, in which students pair up and take on the role of either the tutor or the tutee. During the tutoring session, one student teaches a particular skill to the other student. Each pair is responsible for both practicing the targeted skill and monitoring his or her partner's academic responses. The class divides into two teams, both consisting of peer dyads. Students earn points on the basis of their behavior as a tutor or tutee during a session as well as the number of correct responses or error corrections during a peer-tutoring session. Periodic quizzes maintain individual accountability. Teachers can assign points for correct answers. Instructional guidelines recommend that tutoring sessions last ten to fifteen minutes and occur two to four days during a week.

Despite the instructional recommendations, one must recognize that all research in algebra involving students with disabilities used adolescents or adults and involved primarily students with learning disabilities. Although several authors suggest the need to focus on conceptual understanding as important for developing algebraic understanding (e.g., Ives 2007; Maccini and Ruhl 2000; Witzel 2005; Witzel, Smith, and Brownell 2001), most available research focuses on routinizing procedures that students have difficulty applying problems to. Further, as Foegen (2008) notes, "most of the research has been conducted with relatively simple concepts and problem types in algebra" (p. 76). Although the teaching of procedures can temporarily improve students' uses of them, such instructional techniques can bypass students' natural ways of reasoning and lead to difficulties with long-term retention. Also, much research, including research in algebra, notes a consistent lack of generalization of what has been learned beyond a similar problem type (e.g., Maccini and Hughes 2000).

Supporting All Learners with Algebraic Thinking

Developing algebraic understanding in all students is vital to developing mathematical proficiency (National Mathematics Advisory Panel 2008; RAND 2003). This recommendation is for all students, including those with disabilities. However, working with learners who struggle in mathematics, including algebra, can be challenging. Research focused specifically on this group of students, particularly at the elementary level, is limited.

To assist students to develop a broader understanding of algebraic thinking, we recommend using the various aspects of algebra that Kaput (2008) identified. These recommendations align with NCTM's *Principles and Standards* and *Curriculum Focal Points* and with the *Common Core State Standards* (CCSSO 2010), which can help all students build a foundation of understanding that can facilitate success with algebra in later grades. We also draw on four *Guiding Principles of Instruction* (Morocco 2001) for instructional recommendations for students experiencing mathematical difficulties.

First, use authentic tasks (e.g., real-world tasks, alternative representations of complex ideas) that let students explore ideas and ways of knowing, and explain problems and issues that have relevance and value beyond the school setting. Second, teach cognitive strategies. Students can acquire such strategies through direct or explicit instruction, as previously discussed; however, consider a broader focus when the goal is to help students develop cognitive tools that enable them to work toward understanding. Third, use social mediation, where students serve as intellectual partners to one another. Teachers can implement this principle through using peer-tutoring strategies such as those described previously. However, teachers should also use whole-class mathematical discourse to promote shared ownership, to make thinking visible to other students, and as a means of encouraging a variety of perspectives. Fourth, give students opportunities for constructive conversation with each other that makes their "thinking visible and encourages them to connect, compare, contrast and negotiate different understandings" (p. 9).

Reflections on Teaching Algebra: Questions for Discussion

1. How is learning algebra different from algebraic thinking?

2. How can an emphasis on algebraic thinking help student learning in number and operations?

3. How are the following aspects of algebraic thinking that Kaput (2008) delineated similar to or different from your initial conceptions of algebra?

   - Building generalizations from arithmetic and quantitative reasoning

   - Generalizing patterns toward the idea of function, including determining whether two expressions are equivalent, where functions take on particular values, and whether they satisfy various constraints (building and solving equations)

- Modeling various situations (creating generalizations that characterize the relationship between two quantities)

4. Why is learning algebra so difficult for all students?

5. How can various representations help students with disabilities learn algebraic concepts? What are the strengths and limitations of the various representations that could cause difficulties for student learning of algebraic concepts?

6. How can an awareness of common errors with regard to the learning of formal algebra symbols guide instruction to better meet the needs of students with disabilities?

7. Why is using strategies other than simply slowing the pace of instruction and furnishing manipulatives important for teachers of students with disabilities?

References

Allsopp, David H. "Using Classwide Peer Tutoring to Teach Beginning Algebra Problem-Solving Skills in Heterogeneous Classrooms." *Remedial and Special Education* 18 (November–December 1997): 367–79.

———. "Using Modeling, Manipulatives, and Mnemonics with Eighth-Grade Math Students." *Teaching Exceptional Children* 32 (November–December 1999): 74–81.

Booth, Lesley R. *Algebra: Children's Strategies and Errors.* Windsor, UK: NFER-Nelson, 1984.

Boyer, Carl B., and Uta C. Merzback. "A History of Mathematics." 2nd ed. New York: John Wiley and Sons, 1989.

Carpenter, Thomas P., Megan Loef Franke, and Linda Levi. *Thinking Mathematically: Integrating Arithmetic & Algebra in Elementary School.* Portsmouth, N.H.: Heinemann, 2003.

Carraher, David W., and Analúcia D. Schliemann. "Early Algebra and Algebraic Reasoning." In *Second Handbook of Research on Mathematics Teaching and Learning: A Project of the National Council of Teachers of Mathematics*, edited by Frank K. Lester Jr., vol. II, pp. 669–705. Charlotte, N.C.: Information Age Publishing, 2007.

Cawley, John F., and James H. Miller. "Cross-Sectional Comparisons of the Mathematics Performance of Children with Learning Disabilities: Are We on the Right Track toward Comprehensive Programming?" *Journal of Learning Disabilities* 22 (April 1989): 250–54.

Cawley, John F., Rene Parmar, Teresa E. Foley, Susan Salmon, and Sharmila Roy. "Arithmetic Performance of Students: Implications for Standards and Programming." *Exceptional Children* 67 (Spring 2001): 311–28.

Council of Chief State School Officers (CCSSO). *Common Core State Standards.* Washington, D.C.: CCSSO, 2010. www.corestandards.org/the-standards/mathematics.

Deshler, Donald D., B. Keith Lenz, Janis Bulgren, Jean B. Schumaker, Betsy Davis, Bonnie Grossen, and Janet Marquis. "Adolescents with Disabilities in High School Setting: Student Characteristics and Setting Dynamics." *Learning Disabilities: A Contemporary Journal* 2 (September 2004): 30–48.

Foegen, Anne. "Algebra Progress Monitoring and Interventions for Students with Learning Disabilities." *Learning Disability Quarterly* 31 (March 2008): 65–78.

Gagnon, Joseph Calvin, and Paula Maccini. "Preparing Students with Disabilities for Algebra." *Teaching Exceptional Children* 34 (September–October 2001): 8–15.

Harper, Eon. "Ghosts of Diophantus." *Educational Studies in Mathematics* 18 (February 1987): 75–90.

Healy, Lulu, and Celia Hoyles. "Visual and Symbolic Reasoning in Mathematics: Making Connections with Computers?" *Mathematical Thinking and Learning* 1, no. 1 (1999): 59–84.

Hutchinson, Nancy L. "Effects of Cognitive Strategy Instruction on Algebra Problem Solving of Adolescents with Learning Disabilities." *Learning Disability Quarterly* 16 (Winter 1993): 34–63.

Ives, Bob. "Graphic Organizers Applied to Secondary Algebra Instruction for Students with Learning Disorders." *Learning Disabilities Research and Practice* 22 (May 2007): 110–18.

Ives, Bob, and Cheri Hoy. "Graphic Organizers Applied to Higher-Level Secondary Mathematics." *Learning Disabilities Research and Practice* 18 (February 2003): 36–51.

Kaput, James J. "What Is Algebra? What Is Algebraic Reasoning?" In *Algebra in the Early Grades*, edited by James J. Kaput, David W. Carraher, and Maria L. Blanton, pp. 5–17. Mahwah, N.J.: Lawrence Erlbaum, 2008.

Kenney, Patricia Ann, Judith S. Zawojewski, and Edward A. Silver. "Marcy's Dot Pattern." *Mathematics Teaching in the Middle School* 3 (May 1998): 474–77.

Kieran, Carolyn. "Concepts Associated with the Equality Symbol." *Educational Studies in Mathematics* 12 (August 1981): 317–26.

———. "Learning and Teaching Algebra at the Middle School through College Levels: Building Meaning for Symbols and Their Manipulation." In *Second Handbook of Research on Mathematics Teaching and Learning*, edited by Frank K. Lester Jr., vol. II, pp. 707–62. Charlotte, N.C.: Information Age Publishing, 2007.

Knuth, Eric J., Ana C. Stephens, Nicole M. McNeil, and Martha W. Alibali. "Does Understanding the Equal Sign Matter? Evidence from Solving Equations." *Journal for Research in Mathematics Education* 36 (July 2006): 297–312.

Kortering, Larry J., Laurie U. deBettencourt, and Patricia M. Braziel. "Improving Performance in High School Algebra: What Students with Learning Disabilities Are Saying." *Learning Disability Quarterly* 28 (Summer 2005): 191.

Küchemann, D. "Algebra." In *Children's Understanding of Mathematics: 11–16*, edited by K. M. Hart, pp. 102–19. London: John Murray, 1981.

Lannin, John K., Brian E. Townsend, Nathan Armer, Savanna Green, and Jessica Schneider. "Developing Meaning for Algebraic Representations: Possibilities and Pitfalls." *Mathematics Teaching in the Middle School* 13 (April 2008): 478–83.

Maccini, Paula, and Charles A. Hughes. "Effects of a Problem-Solving Strategy on the Introductory Algebra Performance of Secondary Students with Learning Disabilities." *Learning Disabilities Research and Practice* 15 (Winter 2000): 10–21.

Maccini, Paula, and Kathy L. Ruhl. "Effects of a Graduated Instructional Sequence on the Algebraic Subtraction of Integers by Secondary Students with Learning Disabilities." *Education and Treatment of Children* 23 (November 2000): 465–89.

Maccini, Paula, David McNaughton, and Kathy L. Ruhl. "Algebra Instruction for Students with Learning Disabilities: Implications from a Research Review." *Learning Disability Quarterly* 22 (Spring 1999): 113–26.

Matz, M. "Towards a Computations Theory of Algebraic Competence." *Journal of Mathematical Behavior* 3, no. 1 (1980): 93–166.

Miles, Dorothy D., and Jonathan P. Forcht. "Mathematics Strategies for Secondary Students with Learning Disabilities or Mathematics Deficiencies: A Cognitive Approach." *Intervention School and Clinic* 31 (November 1995): 91–96.

Morocco, Catherine Cobb. "Teaching for Understanding with Students with Disabilities: New Directions for Research on Access to the General Education Curriculum." *Learning Disability Quarterly* 24 (Winter 2001): 5–13.

Nathan, Mitchell J., and Kenneth R. Koedinger. "Teachers' and Researchers' Beliefs about the Development of Algebraic Reasoning." *Journal for Research in Mathematics Education* 31 (March 2000): 168–90.

National Council of Teachers of Mathematics (NCTM). *Principles and Standards for School Mathematics.* Reston, Va.: NCTM, 2000.

———. *Curriculum Focal Points for Prekindergarten through Grade 8 Mathematics: A Quest for Coherence.* Reston, Va.: NCTM, 2006.

National Mathematics Advisory Panel. *Foundations for Success: The Final Report of the National Mathematics Advisory Panel.* Washington, D.C.: U.S. Department of Education, 2008.

Pólya, George. *How to Solve It: A New Aspect of Mathematical Method.* 2nd ed. Garden City, N.Y.: Doubleday, 1957.

Radford, Luis. "Signs and Meanings in Students' Emergent Algebraic Thinking: A Semiotic Analysis." *Education Studies in Mathematics* 42, no. 3 (2000): 237–68.

RAND Mathematics Study Panel. *Mathematical Proficiency for All Students: Toward a Strategic Research and Development Program in Mathematics Education.* Santa Monica, Calif.: RAND Corp., 2003.

Schoenfeld, Alan. "Report of Working Group 1." In *The Algebra Initiative Colloquium,* edited by Carole LaCampagne, vol. II, pp. 11–18. Washington, D.C.: U.S. Department of Education, Office of Educational Research and Improvement, 1995.

Simon, Martin, and Glendon Blume. "Justification in the Mathematics Classroom: A Study of Prospective Elementary Teachers." *Journal of Mathematical Behavior* 15 (March 1996): 3–31.

Sfard, Anna, and Liora Linchevski. "The Gains and the Pitfalls of Reification—The Case of Algebra." *Educational Studies in Mathematics* 26 (March 1994): 191–228.

Stacey, Kaye. "Finding and Using Patterns in Linear Generalising Problems." *Educational Studies in Mathematics* 20 (May 1989): 147–64.

Stacey, Kaye, and Mollie MacGregor. "Curriculum Reform and Approaches to Algebra." In *Perspectives on School Algebra,* edited by R. Sutherland, pp. 141–53. Dordrecht, Netherlands: Kluwer Academic Publishers, 2001.

Steele, Marcee M., and John W. Steele. "Teaching Algebra to Students with Learning Disabilities." *Mathematics Teacher* 96 (December 2003): 622–24.

Swafford, Jane O., and Cynthia W. Langrall. "Grade 6 Students' Preinstructional Use of Equations to Describe and Represent Problem Situations." *Journal for Research in Mathematics Education* 31 (January 2000): 89–112.

Wirt, John, Patrick Rooney, Susan Choy, Stephen Provasnik, Anindita Sen, and Richard Tobin. *The Condition of Education 2004* (No. NCES 2004-077). Washington, D.C.: U.S. Government Printing Office, U.S. Department of Education, National Center for Education Statistics, 2004.

Witzel, Bradley S. "Using CRA to Teach Algebra to Students with Math Difficulties in Inclusive Settings." *Learning Disabilities: A Contemporary Journal* 3 (September 2005): 49–60.

Witzel, Bradley S., Cecil D. Mercer, and David M. Miller. "Teaching Algebra to Students with Learning Difficulties: An Investigation of an Explicit Instruction Model." *Learning Disabilities Research and Practice* 18 (May 2003): 121–31.

Witzel, Bradley S., Paul J. Riccomini, and Elke Schneider. "Implementing CRA with Secondary Students with Learning Disabilities in Mathematics." *Intervention in School and Clinic* 43, no. 5 (2008): 270–76.

Witzel, Bradley S., Stephen W. Smith, and Mary T. Brownell. "How Can I Help Students with Learning Disabilities in Algebra?" *Intervention in School and Clinic* 37 (November 2001): 101–04.

CHAPTER 7

Geometry

Julie Sarama,
Douglas H. Clements,
René S. Parmar,
and René Garrison

The focus of this chapter is geometry and how students with special needs, including learning disabilities, access geometric concepts and relationships. We discuss geometry's role for children with special needs, give an overview of the National Council of Teachers of Mathematics (NCTM) *Principles and Standards for School Mathematics* (NCTM 2000) and *Curriculum Focal Points for Prekindergarten through Grade 8 Mathematics: A Quest for Coherence* (NCTM 2006) for geometry, discuss the difficulties in geometry and spatial sense that students with a variety of disabilities encounter, and offer examples of instructional techniques and modifications to meet the needs of students with disabilities.

Knowledge of geometry concepts is essential in many real-life contexts. It underlies design for everything from microprocessors to large structures. It also forms the basis for understanding and representing objects in the world around us and our geographic and spatial orientation. Although special education has directed considerable effort to enhancing students' number and operations capabilities during the early school years, we feel that concurrently developing geometry concepts and relationships to give students tools necessary to successfully participate in a variety of professions in their future is vital. Spatial and geometric understandings are essential not only in and of themselves but also for the supporting role they play in learning algebra (National Mathematics Advisory Panel 2008). The geometric understandings discussed here reflect the geometry standards within the *Common Core State Standards* (Council of Chief State School Officers [CCSSO] 2010).

Expectations and Needs

What role should geometry play in the mathematics education of children with special needs? Thus far, mathematics curricula for students with disabilities have not sufficiently emphasized and integrated many aspects of geometry. This section explores issues related to focusing on geometry, concluding with an overview of NCTM's standards for geometry as *Principles and Standards for School Mathematics* (NCTM 2000) describes.

Geometry's role for children with special needs

Mathematics interventions for students with disabilities have focused primarily on number and operations, particularly work with whole numbers, with little attention to geometry. In fact, entire books exist about curriculum and instruction for students with various disabilities, discussing topics such as number, arithmetic, ratio, proportion, and algebra, but with no distinct mention of geometry. (We recognize that not as much research on the geometric learning of students with special needs exists [National Mathematics Advisory Panel 2008] but see this as part of the problem.) Research that does mention geometry sometimes devalues it: "Geometry is seen as a relief from working with numbers and not an end in itself" (Shaw 1984, p. 191).

We believe that such perspectives miss the point in four ways. First, the implicit definition of geometry in many mathematics texts and articles is too narrow. Geometry is not just about shape names and geometric proofs. Hans Freudenthal presents a broader view:

> What is geometry? Such questions can be answered on different levels. On the highest, geometry, somehow axiomatically organized is a certain part of mathematics that for some historical reason is called geometry. Mindful of the educational principles I have advocated, I would rather ask what geometry is on the lowest, the bottom level? There can be no doubt what I should then answer—geometry is grasping space. And since it is about the education of children, it is grasping that space in which the child lives, breathes, and moves. The space that the child must learn to know, explore, conquer, in order to live, breathe, and move better in it. (1973, pp. 402–03)

Second, many aspects of geometry involve practical and useful knowledge, not just rote memorization of obscure terminology. Geometry includes navigation and map reading, knowledge of structures used in many trades, and connections to the arts (Thornton, Langrall, and Jones 1997).

Third, geometry activities, especially if consciously designed, develop ideas and skills that go beyond the domain of geometry. They contribute, sometimes greatly, to the development of other areas of mathematics. For example, building geometric and spatial abilities helps students build array models for multiplication and rectangular and circular area models for fractions. Geometric models also contribute to learning about and solving problems in measurement, graphing, ratio and proportion, probability, and algebra. High-achieving students' numerical ability is connected to their spatial and measurement ability. Further, children who are poor achievers in mathematics generally show little growth in geometry. Educators must address geometry and spatial thinking (Stewart, Leeson, and Wright 1997).

Fourth, geometric and spatial tasks can reveal, and often build on, unsuspected strengths of students with special needs. They can capitalize on student strengths in drawing or manipulating forms, thus offering alternatives for students with language and communication difficulties. They lend themselves to small-group project approaches and immediate, real-world applications. For most students, such tasks are motivating. They often create nonthreatening, appealing invitations back into the world of mathematics. Other researchers similarly state that geometry is an important area of problem solving, in which children with disabilities can build on their strengths (Swanson 1993; Thornton, Langrall, and Jones 1997; Wansart 1990).

Geometry expectations and focal points

The *Principles and Standards* key aspects of geometry (pp. 41–43) that describe what pre-K–12 instructional programs should enable students to do are as follows:

1. "Analyze characteristics and properties of two- and three-dimensional geometric shapes and develop mathematical arguments about geometric relationships." Students should be able to identify and name, but also describe and reason about, two- and three-dimensional shapes. As they discuss shapes, their properties, and topics such as similarity and congruence, students learn to reason and explain themselves logically.

2. "Specify locations and describe spatial relationships using coordinate geometry and other representational systems." From the earliest age, all children need to learn to understand and converse about relative position, such as *above*, *behind*, *near*, *far*, and *between*. Understanding one-dimensional space, such as on a number line, can also support young students' learning of whole numbers and fractions. Eventually, they can learn about paths and maps and use coordinates to name and locate places on such maps (Sarama et al. 2003). Coordinates also are useful in understanding functions and other algebraic concepts. Understanding grids can also support understanding multiplication (Battista et al. 1998).

3. "Apply transformations and use symmetry to analyze mathematical situations." Students can build on intuitions about moving objects in space to explore motions such as slides, flips, and turns until they understand the properties and effects of geometric transformations. They can explore symmetry by using mirrors, paper folding, and tracing and can continue to make connections with other sciences and the arts.

4. "Use visualization, spatial reasoning, and geometric modeling to solve problems." From the earliest years, students should develop their ability to understand, analyze, and represent different perspectives or views. They should be able to translate between two-and three-dimensional shapes and their representations. Such visualization skills can assist students in many different topics, subjects, and out-of-school activities.

Geometry includes a wide range of skills—from recognizing simple shapes to creating complex proofs, from understanding *above* and *beside* to navigating complex spaces with maps and graphing algebraic functions. Through learning geometric principles, students develop reasoning and justification skills, as well as learn to interpret and describe physical environments (NCTM 2000).

NCTM's *Curriculum Focal Points* (2006) describes geometric areas of focus grade by grade. This is an important guide for what is most essential for children to learn but is not an exhaustive list. Figure 7.1 summarizes the main Focal Points for geometry (pre-K–8):

Pre-K

Geometry: Identifying shapes and describing spatial relationships

Children develop spatial reasoning by working from two perspectives on space as they examine the shapes of objects and inspect their relative positions. They find shapes in their environments and describe them in their own words. They build pictures and designs by combining two- and three-dimensional shapes, and they solve such problems as deciding which piece will fit into a space in a puzzle. They discuss the relative positions of objects with vocabulary such as "above," "below," and "next to."

Kindergarten

Geometry: Describing shapes and space

Children interpret the physical world with geometric ideas (e.g., shape, orientation, spatial relations) and describe it with corresponding vocabulary. They identify, name, and describe a variety of shapes, such as squares, triangles, circles, rectangles, (regular) hexagons, and (isosceles) trapezoids presented in a variety of ways (e.g., with different sizes or orientations), as well as such three-dimensional shapes as spheres, cubes, and cylinders. They use basic shapes and spatial reasoning to model objects in their environment and to construct more complex shapes.

Grade 1

Geometry: Composing and decomposing geometric shapes

Children compose and decompose plane and solid figures (e.g., by putting two congruent isosceles triangles together to make a rhombus), thus building an understanding of part–whole relationships as well as the properties of the original and composite shapes. As they combine figures, they recognize them from different perspectives and orientations, describe their geometric attributes and properties, and determine how they are alike and different, in the process developing a background for measurement and initial understandings of such properties as congruence and symmetry.

Grade 2

Geometry: Connections to the Focal Points

Children estimate, measure, and compute lengths as they solve problems involving data, space, and movement through space. By composing and decomposing two-dimensional shapes (intentionally substituting arrangements of smaller shapes for larger shapes or substituting larger shapes for many smaller shapes), they use geometric knowledge and spatial reasoning to develop foundations for understanding area, fractions, and proportions.

Grade 3

Geometry: Describing and analyzing properties of two-dimensional shapes

Students describe, analyze, compare, and classify two-dimensional shapes by their sides and angles and connect these attributes to definitions of shapes. Students investigate,

---

Fig. 7.1. Curriculum Focal Points for geometry and spatial thinking

describe, and reason about decomposing, combining, and transforming polygons to make other polygons. Through building, drawing, and analyzing two-dimensional shapes, students understand attributes and properties of two-dimensional space and the use of those attributes and properties in solving problems, including applications involving congruence and symmetry.

Grade 4
Geometry: Connections to the Focal Points

Students extend their understanding of properties of two-dimensional shapes as they find the areas of polygons. They build on their earlier work with symmetry and congruence in grade 3 to encompass transformations, including those that produce line and rotational symmetry. By using transformations to design and analyze simple tilings and tessellations, students deepen their understanding of two-dimensional space.

Grade 5
Geometry and Measurement and Algebra: Describing three-dimensional shapes and analyzing their properties, including volume and surface area

Students relate two-dimensional shapes to three-dimensional shapes and analyze properties of polyhedral solids, describing them by the number of edges, faces, or vertices as well as the types of faces. Students recognize volume as an attribute of three-dimensional space. They understand that they can quantify volume by finding the total number of same-sized units of volume that they need to fill the space without gaps or overlaps. They understand that a cube that is 1 unit on an edge is the standard unit for measuring volume. They select appropriate units, strategies, and tools for solving problems that involve estimating or measuring volume. They decompose three-dimensional shapes and find surface areas and volumes of prisms. As they work with surface area, they find and justify relationships among the formulas for the areas of different polygons. They measure necessary attributes of shapes to use area formulas to solve problems.

Grade 6
Geometry: Connections to the Focal Points

Problems that involve areas and volumes, calling on students to find areas or volumes from lengths or to find lengths from volumes or areas and lengths, are especially appropriate. These problems extend the students' work in grade 5 on area and volume and provide a context for applying new work with equations.

Grade 7
Measurement and Geometry and Algebra: Developing an understanding of and using formulas to determine surface areas and volumes of three-dimensional shapes

By decomposing two- and three-dimensional shapes into smaller, component shapes, students find surface areas and develop and justify formulas for the surface areas and volumes of prisms and cylinders. As students decompose prisms and cylinders by slicing them, they develop and understand formulas for their volumes (Volume = Area of base × Height). They apply these formulas in problem solving to determine volumes of prisms and cylinders. Students see that the formula for the area of a circle is plausible by decomposing a circle into a number of wedges and rearranging them into a shape that approximates

Fig. 7.1. Curriculum Focal Points for geometry and spatial thinking—*Continues*

a parallelogram. They select appropriate two- and three-dimensional shapes to model real-world situations and solve a variety of problems (including multistep problems) involving surface areas, areas and circumferences of circles, and volumes of prisms and cylinders.

Geometry: Connections to the Focal Points

Students connect their work on proportionality with their work on area and volume by investigating similar objects. They understand that if a scale factor describes how corresponding lengths in two similar objects are related, then the square of the scale factor describes how corresponding areas are related, and the cube of the scale factor describes how corresponding volumes are related. Students apply their work on proportionality to measurement in different contexts, including converting among different units of measurement to solve problems involving rates such as motion at a constant speed. They also apply proportionality when they work with the circumference, radius, and diameter of a circle; when they find the area of a sector of a circle; and when they make scale drawings.

Grade 8
Geometry and Measurement: Analyzing two- and three-dimensional space and figures by using distance and angle

Students use fundamental facts about distance and angles to describe and analyze figures and situations in two- and three-dimensional space and to solve problems, including those with multiple steps. They prove that particular configurations of lines give rise to similar triangles because of the congruent angles created when a transversal cuts parallel lines. Students apply this reasoning about similar triangles to solve a variety of problems, including those that ask them to find heights and distances. They use facts about the angles that are created when a transversal cuts parallel lines to explain why the sum of the measures of the angles in a triangle is 180 degrees, and they apply this fact about triangles to find unknown measures of angles. Students explain why the Pythagorean theorem is valid by using a variety of methods—for example, by decomposing a square in two different ways. They apply the Pythagorean theorem to find distances between points in the Cartesian coordinate plane to measure lengths and analyze polygons and polyhedra.

Geometry: Connections to the Focal Points

Given a line in a coordinate plane, students understand that all "slope triangles"—triangles created by a vertical "rise" line segment (showing the change in y), a horizontal "run" line segment (showing the change in x), and a segment of the line itself—are similar. They also understand the relationship of these similar triangles to the constant slope of a line.

Fig. 7.1. Curriculum Focal Points for geometry and spatial thinking—*Continued*

Students with Special Needs, Geometry, and Spatial Sense

Children with special needs vary greatly in their relationship with geometry. For example, the nature of this domain requires special consideration for those with visual impairments.

Students with learning disabilities

In this section, we overview the learning needs of children with a variety of learning disabilities and draw implications for learning geometry and developing spatial sense.

Spatial organization and number

Some students have difficulty with spatial organization across a wide range of tasks. Students who experience difficulty with visual processing will probably also experience difficulties with mathematics, which requires the ability to visualize numbers and geometric concepts (Sousa 2001). Students with certain mathematics learning disabilities may struggle with spatial relationships, visual–motor and visual perception, and a poor sense of direction (Lerner 1997). Altering the orientation of geometric shapes may compound difficulties with visual perception (Shaw and Durden 1998), such as understanding that a square is still a square when turned on its side. Some research suggests that ideas about number and arithmetic are closely connected to spatial competencies. (See the sidebar "Research on Spatial Sense and Arithmetic.")

For students with difficulty with spatial organization, then, remediation may be important to their overall mathematics development. These students may often have difficulties with visual processing, including spatial perception, figure–ground relationships, and visual discrimination. Ordering diagrams and notes on a page, or objects in their desk, particularly challenges some students.

Some of these students may have strong verbal abilities, and talking about visual patterns, arrangements, and structures (van Hiele 1986) may help them build from their competencies. For example, if encouraged to describe properties of geometric shapes, they can check new figures against their verbal criteria, thereby transferring visual to verbal learning. For example, they can learn that *octagon* means "eight angles," relating *oct-* to octopus and octave (eight diatonic degrees in a scale). For *parallelogram*, they might learn not only the root "parallel" but also properties such as two pairs of equal and parallel sides, among others. For practical, everyday tasks, labeled containers, notebooks and sections of notebooks, and other concrete organizers can be helpful.

Focusing attention

Some students have difficulty focusing their attention on the relevant aspects of learning materials. Strategies that may help include having students close to the board during presentations and discussions, using larger-than-normal size for diagrams and print materials, using colors or clear areas to separate important aspects of problems, allowing enough time for students to process visual material, and using manipulatives. Teachers should guide students to adopt these strategies for their independent use (Sliva 2004).

Finally, such students may also "turn around" their lack of interest and engagement with spatial tasks if they engage in computer graphics. One can create fascinating visual forms with coordinates or turtle graphics (as in the LOGO computer language). Also, software such as the Geometer's Sketchpad allows students to construct and manipulate geometric shapes and angles and to explore their mathematical properties. For others, Web design with a variety of graphic tools might be a path to comfort and active involvement with spatial organization.

Research on Spatial Sense and Arithmetic

Some researchers believe that ideas about number are originally based in, or at least connected to, spatial quantities (Mix, Huttenlocher, and Levine 2002; Spiers 1987). So, they believe that spatial deficits in childhood may be especially detrimental to the development of basic numerical skills (Semrud-Clikeman and Hynd 1990; Spiers 1987).

Students with learning disabilities may show a spatial type of arithmetic deficit. For example, children with selective arithmetic deficits performed significantly worse on the WISC Performance Scale than students with both arithmetic and language deficits.

Research also supports a connection between spatial and numerical abilities. For example, nine-year-old students with arithmetic difficulties have normal working memory but are impaired on spatial working memory, as well as some aspects of executive processing (McLean and Hitch 1999). Consistent results manifest in children with Williams syndrome, a rare genetic disorder. Children with Williams syndrome have relatively strong language, music, social, and face-processing abilities but are impaired on visual–spatial cognition. For example, they have great difficulty copying geometric shapes, even line segments and triangles, or copying a design with color blocks (Bellugi et al. 2000). They can learn basic reading and spelling tasks. They also have considerable difficulties with mathematics, especially arithmetic (Bellugi et al. 2000; Howlin, Davies, and Udwin 1998). Thus, these children, too, have specific impairments in both spatial geometric and arithmetic skills.

In a similar vein, researchers report an association between arithmetic deficits and visual–spatial abilities. One study involved three groups of students with various learning disabilities. Rourke and Finlayson (1978) evaluated brain functioning in children with low performances in (1) reading, spelling, and arithmetic; (2) only in spelling and reading; and (3) only in arithmetic. Only the last group, average or above in reading and spelling but deficient in arithmetic, showed a dysfunctional right, but not left, cerebral hemisphere, indicated by low performance on visual–spatial tasks. Thus, low visual and spatial organization and integration may limit children's arithmetic abilities (Rourke and Finlayson 1978). This is not to say that all arithmetic disabilities are related to a dysfunctional right cerebral hemisphere. Some arithmetic learning disabilities appear to be related to difficulties with verbal deficiencies and others to visual–spatial deficiencies (Rourke and Conway 1997). The last group of children, those who have low achievement in arithmetic but average reading scores, often perform poorly on measurements of spatial abilities and on timed, but not untimed, arithmetic tests (Geary, Hoard, and Hamson 1999; Rourke and Finlayson 1978).

Such relationships may be true more for boys than for girls (Share, Moffitt, and Silva 1988). Some studies have shown a connection between right-hemisphere dysfunction and low arithmetic abilities (Gross-Tsur et al. 1995; Tranel et al. 1987). Damage to the parietal lobe can also cause deficits in mathematical ability (Sousa 2001).

Teachers must monitor a particular pattern: nonverbal learning disabilities, such as in the children who had lower visual and spatial skills and low arithmetic abilities but well-developed word recognition and spelling (Rourke and Finlayson 1978). This pattern, unlike other learning disability patterns, associates with increased risk of socioemotional disturbance, psychopathology, and greater discrepancies between the children's assets and deficits (Rourke 1991).

Connections between spatial and numerical abilities may exist for most children. Deeper conceptual connections may be important as well. For example, understanding area and volume requires children to learn spatial structuring. *Spatial structuring* is the mental operation of constructing an organization or form for an object or set of objects in space, such as organizing squares into rows and columns (Sarama et al. 2003). That same spatial structuring is a major mental representation for multiplication. So, children who have difficulties with spatial abilities may also show impaired ability to visualize arithmetic tasks, from "seeing" 7 as "5 and 2" in a ten frame to creating a mental image of rows and columns in multiplication, area, or volume tasks (Battista and Clements 1996; Battista et al. 1998). Helping children, especially those with less-developed spatial abilities, develop spatial structuring is therefore important for several areas of mathematics. Doing so is particularly important for students, who need to develop concepts and adaptive reasoning skills to learn number and operation concepts, especially with full understanding (Rourke and Conway 1997).

Spatial and geometric knowledge and levels of geometric thinking

Students with learning disabilities may not perceive a shape as a complete and integrated entity. For example, a triangle may appear to them as three separate lines, as a rhombus, or even as an undifferentiated closed shape (Lerner 1997). Developmental teaching is even more important for children with learning disabilities, as well as children with other special needs. Teachers should know the developmental sequences through which children pass as they learn geometric ideas.

According to the theory of Pierre van Hiele and Dina van Hiele-Geldof, students progress through levels of thought in geometry (van Hiele 1986; van Hiele-Geldof 1984). See the sidebar "The van Hiele Theory" for descriptions.

At level 0, prerecognition, children cannot reliably distinguish or recognize shapes. These students need many experiences finding, naming, and discussing shapes, including the faces of real-world objects and manipulatives. To grasp spatial relations, students could act out and talk about many meanings for words such as *between*—for example, sitting in a circle, who is between two other children. "I'm thinking of a person who is between Mary and John." Continue to expand puzzles such as these to notions of between in daily work ("Get the chair between the sink and the wall") and then to other descriptions of geometric relations and even shapes ("I'm thinking of a shape that has four sides . . ."). Students might also fold and construct shapes from sticks or copy shapes on geoboards.

At the visual level, level 1, children can recognize shapes only as wholes. They do not think about the attributes, or properties, of shapes. These students should learn to match, compare, and describe two-dimensional shapes and relate plane figures to faces of real objects. They might use "feeling boxes" to explore and identify shapes by touch. They can learn attributes of shapes by playing "one of these things is not like the others" with open and closed containers, open and closed shapes, shapes that have all straight sides and shapes that have one or more sides that are not straight, shapes that are triangles and shapes that are not, and so forth. Students might measure, color, fold, and cut to identify properties of figures, such as folding a square or rhombus in various ways to determine symmetries or determine the equality of angles and sides. They could sort shapes by attributes or play "guess my shape" from property clues. They could work with the class to make up definitions of shapes.

At level 2, descriptive/analytic, students recognize and characterize shapes by their properties. For instance, a student might think of a square as a figure with four equal sides and four right angles. To move to level 3, students might construct shapes with manipulatives such as D-Stix or computer-based tools such as LOGO and geometric construction software. They could play games with "property cards" in which they have to change a figure on a geoboard to have the property described on the card they choose. They could work in groups to define figures in one or more ways. For example, a rectangle might be "a shape with two pairs of equal and parallel sides and all right angles," "a quadrilateral with all right angles," "a parallelogram with at least one right angle," and many others.

At level 3, abstract/relational, students can form abstract definitions, distinguish between necessary and sufficient sets of conditions for a concept, and understand and sometimes even produce logical arguments in the geometric domain. Teachers might ask students to deduce that in any quadrilateral the sum of the angles must be 360° because any quadrilateral consists of two triangles, all of whose angles sum to 180°.

In summary, the van Hiele learning theory offers a general framework for curriculum and teaching. It also reminds us that students think about geometry in quite different ways as well as serving as a framework that helps us understand students' variegated notions. We cannot ignore that most U.S. students are not developing through the levels at all (Fuys, Geddes, and Tischler, unpublished data), but such development is possible with better curriculum and teaching. For example, most textbooks do not require students to develop higher levels of thinking through the grades (Fuys, Geddes, and Tischler 1988).

For most students with learning disabilities, making connections to other subject areas and other aspects of their lives is even more important. For example, students might find parallel and perpendicular lines in alphabet letters. They could explore origami and similar constructions such as paper pop-up cards and books. Other situations, such as shadows and "lines of vision," are also appropriate and mathematically fruitful (Gravemeijer 1990).

Many students with learning disabilities have strong geometric and spatial sense. They can build, fix, and create with blocks, Legos, tools, and art media. In the traditional U.S. educational system, they rarely get the opportunity to show, much less develop, these strengths in mathematics. Further, some of them are not as talented or interested in numbers and verbal mathematics. A strong geometry and measurement component helps engage these children in mathematics and can serve as a springboard, both conceptually and motivationally, to other mathematical domains. Some of the vignettes in "Areas of Focus and Instructional Activities" will illustrate the work of such students.

▼

The van Hiele Theory

The van Hiele theory consists of four levels.

Level 0: Prerecognition. Children do not reliably distinguish circles, triangles, and squares from nonexemplars of those classes and appear unable to form reliable mental images of these shapes (Clements et al. 1999).

Level 1: Visual. Students can recognize shapes only as wholes and cannot form mental images of them. They do not think about the attributes, or properties, of shapes. For instance, in classifying quadrilaterals, students at this level included imprecise visual qualities and irrelevant attributes (e.g., orientation) in describing the shapes while omitting relevant attributes. Students start to achieve this level in the early years but may still be at this level even in middle school and, for some, high school.

Level 2: Descriptive/analytic. Students recognize and characterize shapes by their properties. Observing, measuring, drawing, and model making establish properties experimentally. Students discover that some combinations of properties signal a class of figures and some do not; thus, the seeds of geometric implication are planted. Students at this level do not, however, see relationships between classes of figures (e.g., a student might believe that a figure is not a rectangle because it is a square). One girl at this level said that rectangles have "two sides equal and parallel to each other. Two longer sides are equal and parallel to each other, and they connect at 90 degrees" (Burger and Shaughnessy 1986, p. 39). Squares were not included. Many students do not reach this level, even in high school.

Level 3: Abstract/relational. Students can form abstract definitions, distinguish between necessary and sufficient sets of conditions for a concept, and understand and sometimes even present logical arguments in the geometric domain. They can classify figures hierarchically (by ordering their properties) and give informal arguments to justify their classifications (e.g., a square is identified as a rhombus because one can think of it as a rhombus with some extra properties). They can discover properties of classes of figures by informal deduction. As students discover such properties, they feel a need to organize the properties. One property can signal other properties, so definitions can be seen not merely as descriptions but also as a way of logically organizing properties. It becomes clear why, for example, a square is a rectangle. This logical organization of ideas is the first manifestation of true deduction. The students still, however, do not grasp that logical deduction is the method for establishing geometric truths.

Level 4: Formal deduction. Students can establish theorems within an axiomatic system. They recognize the difference among undefined terms, definitions, axioms, and theorems. They can construct original proofs: they can produce a sequence of statements that logically justifies a conclusion as a consequence of the "givens."

Students with autism spectrum disorders

Students with autism spectrum disorders (ASDs) exhibit deficits in communication and social interaction, language impairment, and abnormal behavior such as repetitive acts. ASDs are often associated with some intellectual deficits. However, students with ASDs sometimes excel at geometry, spatial skills, drawing, and computer programming. About 10 percent of children with ASDs exhibit savant (exceptional) abilities, often spatial. Teachers can then use geometry as an area of strength and geometry activities as a way to encourage them to communicate with others.

To ensure that they are perceiving materials, however, students with ASDs should have eye examinations and, as necessary, corrective lenses and visual management training. For many children with good or correct vision, manipulatives and pictures can aid learning of most topics in geometry, number, and other areas, because children with ASDs are often visually oriented. Breaking down what might have been a long verbal explanation or set of directions, and supporting it with pictures, is usually helpful. Processes, too, benefit from dramatizations. For example, teachers might turn a shape slowly when illustrating geometric motion, or a teacher of young children might move a toy airplane up when explaining the concept *up*.

Teachers should find and use the often intense interests that characterize most children with ASDs to motivate them to study geometry and spatial structures. For example, if they enjoy construction, they might study how triangles are used in bridges.

Students with visual impairments

What spatial and geometric skills might students with visual impairments have? Even children blind from birth can infer paths that they have not learned or experienced. Kelli, for example, a 2.5-year-old child who was blind, could figure out what paths to walk between two new pairs of locations after moving between other pairs (Landau, Gleitman, and Spelke 1981). However, this does not mean that spatial sense is innate. Students still have to build it. Kelli probably built her abilities on other available senses, such as her internal senses of movement. Further, children who are blind or visually impaired perform less accurately getting to the final position than age-matched sighted, but blindfolded, children in spatial-inference tasks. This finding is noteworthy because the blindfolding created an artificial task for sighted children (Morrongiello et al. 1995). People blind from birth often have difficulties with spatial tasks, particularly with accuracy in encoding distance and angle increasing with distance between objects (Arditi, Holtzman, and Kosslyn 1988). They tend to represent routes as a sequence of landmarks, rather than having an overall path or two-dimensional representation, the formation of which may require simultaneously experiencing multiple locations (Iverson and Goldin-Meadow 1997). Thus, at least some visual experience appears important for full development of spatial knowledge (Morrongiello et al. 1995; Newcombe and Huttenlocher 2000).

Therefore, all students can build up spatial sense and geometric notions. Spatial knowledge is spatial, not visual. Even children blind from birth are aware of spatial relationships. By age three, they begin to learn about spatial characteristics of certain visual language (Landau 1988). They can learn from spatial–kinesthetic (movement) practice (Millar and Ittyerah 1992). They perform many aspects of spatial tasks similarly to blindfolded sighted children (Morrongiello et al. 1995). Second, visual input is important, but people can construct spatial relations without it (Morrongiello et al. 1995). People who are blind can learn to discriminate the size of objects, or their shape (circle, triangle, and square), with 80 percent accuracy by distinguishing echoes (Rice 1967, as cited in Gibson 1969). They can certainly do so through tactile explorations. For example, blind students have successfully learned to seriate lengths (Lebron-Rodriguez and Pasnak 1977). Primary-grade students can develop the ability to compare rectangular areas by tactile scanning of the two dimensions (Mullet and Miroux 1996).

However, the more severe the visual impairment, the more teachers need to make sure that they give students additional activities that build on their experiences with moving their bodies and feeling objects. Students with low vision can follow activities for sighted students, but with enlarged print, visuals, and manipulatives. Sometimes, using devices to assist low-vision students facilitates their geometry learning.

Using real objects and manipulative solids to represent two- and three-dimensional objects is crucial for all students with visual impairments. One can represent two-dimensional objects in tactile form on a two-dimensional plane adequately, but take care that the entire presentation is not too complex. For example, *Let's Learn Shapes with Shapely-CAL* presents tactile representations of common shapes (Keller and Goldberg 1997).

However, two-dimensional tactile representations are *not* adequate for representing three-dimensional objects. Detailed guidance and elaboration of the students' experiences with such objects is important. This task is labor intensive but an important part of the educational experience for severely visually impaired children. Teachers should make sure that the students explore all components of the object and reflect on how these components relate to each other. Students can explore and describe a three-dimensional solid, reconstruct a solid made of components (such as with Googooplex), and construct a cube given only one edge (e.g., with D-Stix).

Students with hearing impairments

Research with deaf students indicated that both teachers and students often did not have substantial experience with geometry (Mason 1995). Language, however, did play an important role. For example, the American Sign Language sign for *triangle* is roughly equilateral or isosceles. After an eight-day geometry unit, many students finger spelled "t-r-i-a-n-g-l-e" instead of using signs, which may indicate a differentiation in their minds between their new definition of the word *triangle* and what they had previously associated with the sign for *triangle*. With richer learning experiences, a more varied mathematical vocabulary, and exposure to a wide range of geometry concepts, students can succeed and grow in learning geometry (Mason 1995).

Students with limited English proficiency

Given the sometimes confusing vocabulary in geometry education, students with limited English proficiency (LEP) require special attention. One study showed that English-proficient (EP) students and students with LEP can work together using computers to construct the concepts of reflection and rotation. Students experiencing the dynamic computer environment significantly outperformed students experiencing a traditional instructional environment on content measures of the concepts of reflection and rotation as well as on measures of two-dimensional visualization ability. LEP students did not perform statistically significantly differently from their EP peers on tests when experiencing the same instructional environments (Dixon 1995).

Considerations for children at risk

Although they are not the focus of this book, we should not neglect children from low-resource communities, who are often at risk for school failure in mathematics. From the preschool years, educators should give attention to enriching experiences (Clements, Sarama, and DiBiase 2004). Abilities such as classifying (the simple oddity problem "which one is not like the others?") and seriating, or ordering, by length are essential to mathematics learning. Children who lack these spatial–geometric abilities are particularly at risk. Direct instruction on these abilities has led to lasting gains in mathematics achievement (Pasnak et al. 1996).

Direct teaching is not the only, nor necessarily best, way to develop these abilities. Physical and spatial activities for young children help form the general logical–mathematical foundation for mathematics learning (Kamii, Rummelsburg, and Kari 2004). For example, playing pick-up sticks helps children seriate by length to find the next stick, and stacking blocks involves experience with shapes and the logic of balancing and classifying. Another approach is to use gamelike activities that involve these abilities in the context of number activities (Clements 1984).

Teachers and Teaching

Teaching teachers to better serve students with disabilities usually benefits all students (Peterson and Hittie 2003). Students in classrooms of teachers who have engaged in professional development in working with children with special needs outperform other students by more than a full grade level (Wenglinsky 2000). Why? Instructional strategies and approaches that help students with disabilities also help other students. Learning about how different students approach and solve mathematics problems will help teachers deepen their content knowledge and apply it more flexibly, vital competencies associated with teaching excellence in mathematics. Learning how to teach geometry to students with visual impairments, teachers will discover strategies for teaching spatial concepts that are not visual. Exploring a well-known concept in a new way (e.g., through auditory or tactile means) can uncover new, previously unnoticed aspects and relationships of this concept. In the following we illustrate three broad categories of instructional approaches for students with special needs.

Teaching geometry by using a direct instruction approach

The general perspective on mathematics instruction—involving direct teacher explanations; strategy instruction; relevant practice with worksheets, peers, or computer-assisted instruction; formative assessment; and dynamic feedback and reinforcement—represents a currently well-supported view of instruction in special education (Mastropieri et al. 1991). Direct-instruction approaches are particularly suited to teaching specific skill sets and in helping students develop fluency and competence once they have understood concepts.

Direct instruction is highly organized and carefully sequenced. Teachers determine measurable objectives, plan the teaching through task analysis, determine students' mastery of prerequisite skills, instruct explicitly, and plan for continual testing (Lerner 1997). They must also give adequate time for learning and practice to master skills. For example, one could teach the concept of triangle as figure 7.2 shows.

General Teaching Sequence	Illustration: Teaching "Triangle"
Session 1	
1. Describe the defining attributes of the shape and illustrate them with examples and nonexamples.	Show shapes including triangles and other shapes, such as four-sided shapes (quadrilaterals), including shapes that *look like* triangles but are not. Say that a triangle has *three straight sides* (run your fingers along the sides) and *three angles* (touch the corners, or vertices, as you count them) and is closed—no "gaps."
2. Draw each attribute (if necessary, but connecting figure's outline with dots).	Have children trace various triangles with their fingers, discussing the three sides and the three angles.

Fig. 7.2. Direct instruction for the concept of triangle (adapted from McMurray et al. 1977)—*Continues*

3. Show shapes and have children indicate whether each has a particular defining attribute. Give feedback immediately.	Again show the variety of shapes. Ask children to tell whether each is a triangle and defend their decision. Make sure they mention all the attributes: *three straight sides, three angles,* and *closed*.
Session 2	
1. Review all session 1 work.	Review the preceding.
2. Children draw or build examples of the shape with manipulatives.	Have children draw or make triangles. Ask them how they know they are triangles.
3. Show paired shapes, one with a defining attribute and one without it. Have children identify which is an example of the shape and which is not, explaining their answer.	Show a "fooler" (shape that looks like a triangle but is not), such as a chevron, ⋀ , next to a triangle. Ask why one is and one is not a triangle. Do this with all the attributes—for example, a triangular fooler, ⟋⟍ , with sides that are not straight, next to a triangle that does have straight sides, or an open shape, ⟋⟍ , next to an actual triangle, ◁ .
4. Show figures and model, asking whether all the defining attributes are present. Have students do the same for new figures.	"Is this a shape that has three straight sides *and* three angles *and* is closed? Yes, so it is a triangle!" Have children ask these questions themselves to decide on more triangles and foolers. [*Math note:* Triangles are three-sided polygons, and to be a polygon, a figure must be a plane (flat) figure *and* be closed *and* be simple (no crossed lines) *and* consist only of straight sides. We assume these characteristics, but if they emerge in conversations, teachers can add them to the list of necessary attributes.]
Repeat in session 2 as necessary.	

Fig. 7.2. Direct instruction for the concept of triangle (adapted from McMurray et al. 1977)—*Continued*

Teaching geometry by using a constructivist approach

An approach based on the van Hiele theory progresses through five phases in moving students from one level of thinking to the next. In phase 1, information, the teacher places ideas at the student's disposal. In phase 2, guided orientation, students actively explore objects (e.g., folding, measuring) to encounter the principle connections of the network of conceptual relations that will form. In phase 3, explicitation, teachers guide students to become explicitly aware of their geometric conceptualizations, describe these conceptualizations in their own language, and learn traditional mathematical language. In phase 4, free orientation, students solve problems that require synthesizing and using those concepts and relations. The teacher's role includes selecting appropriate materials and geometric problems (with multiple solution paths); giving instructions to permit various performances and to encourage students to reflect and elaborate on these problems and their solutions; and introducing terms, concepts, and relevant problem-solving processes as needed. In phase 5, integration, teachers encourage students to reflect on and to consolidate their geometric knowledge, with more emphasis on using mathematical structures as a framework for consolidation and, eventually, to place these consolidated ideas in the structural organization of formal mathematics. At the completion of phase 5, students attain a new level of thought for the topic. The explicitation and integration phases sharply direct the learner's attention. The van Hiele approach has not been tested with children with special needs; thus, teachers need to consider each phase in relation to students' characteristics. For example, children with learning disabilities solve problems differently, often using less metacognitive knowledge (Swanson 1993). Teachers could scaffold problem-solving activities with this knowledge in mind.

The van Hiele approach implies that building students' visual-level thinking requires much beyond naming shapes. They might make shadows with shapes and identify shapes in different contexts, all the while describing their experiences. Especially at the early levels, children should manipulate concrete geometric shapes and materials so that they can work out geometric shapes on their own. They might combine, fold, and create shapes, or copy shapes on geoboards, by drawing or by tracing. Children who are ready to explore level 2 can investigate the parts and attributes, or properties, of shapes. They might measure, color, fold, or cut to identify properties of figures. For example, children could fold a square to figure out equality of sides or angles or to find symmetry (mirror) lines. They might sort shapes by their attributes (all those with a square corner here) or play "guess my shape" from attribute clues. Imprecise language also plagues students' work in geometry (Burger and Shaughnessy 1986; Fuys, Geddes, and Tischler 1988). Instruction should carefully distinguish common usage from mathematical usage. Teachers need to remember that children's concepts that underlie language may be vastly different from what the teachers think. Thus, when mathematical language appears too early and when the teacher does not use everyday speech as a point of reference, students learn mathematical language without concomitant mathematical understanding (van Hiele-Geldof 1984).

Cognitive strategy instruction

Educators have successfully implemented approaches using cognitive strategy instruction with students with disabilities in arithmetic problem solving but have not directly applied them in geometry. These approaches include modeling; demonstrating; giving specific feedback; teaching cognitive and mathematical strategies, as well as mnemonic, or memory, strategies; cognitive behavior modification; peer mediation; and computer-assisted instruction (Mastropieri, Scruggs, and Shiah 1991).

Mastropieri, Scruggs, and Shiah (1991) cautioned that the *Principles and Standards* innovative approaches have not been adequately tried and tested and asked whether such approaches make too many demands on the learner or whether special educators have not completely understood them. Other researchers believe that innovative approaches have been underused, misused, and misunderstood (Grobecker 1999). We need more research to determine the advantages and disadvantages of these approaches, particularly for learners with special needs. Meanwhile, educators should be aware that research supports all the approaches (direct instruction, constructivist, cognitive strategy instruction), if conducted well, at least in promoting certain aspects of mathematical development.

Other general teaching guidelines

Research also has identified several specific instructional guidelines for children with special needs. For example, a variety of instructional materials is beneficial in meeting the needs of all students (Parmar and Cawley 1997). Pictures give many students an immediate, intuitive grasp of ideas. However, pictures need to vary sufficiently, or students develop limited ideas. For example, students come to believe that all triangles are isosceles and have a horizontal base. Also, teachers must modify pictures for the visually impaired (Keller and Goldberg 1997). Furthermore, pictures are usually not better than manipulatives. Students who use manipulatives in their mathematics classes usually outperform those who don't (Driscoll 1983; Greabell 1978; Raphael and Wahlstrom 1989; Sowell 1989; Suydam 1986). Manipulatives can help students with learning disabilities learn mathematics meaningfully. Manipulatives do not necessarily have to be physical objects. Computer manipulatives can offer students meaningful representations better structured to support learning. Such computer representations may be more manageable, clear, flexible, and extensible than physical manipulatives (Clements and McMillen 1996). Students who use physical and software manipulatives demonstrate a greater mathematical sophistication than do control-group students who use physical manipulatives only (Olson 1988). Good manipulatives can help students connect various pieces of knowledge. For example, computer software can dynamically connect pictured objects, such as a rectangle, to symbolic representations, such as the measure of its sides and angles. More guidelines from research follow.

1. Understand students' development of the *Curriculum Focal Points* (NCTM 2006), the National Mathematics Advisory Panel's (2008) critical foundations, or a state or school district's "big ideas" of geometry, and use scaffolding to help students develop more sophisticated ideas and strategies (Kameenui and Carnine 1998; Swanson and Hoskyn 1998), encouraging student exploration and invention at every stage.

2. Keep expectations reasonable but not low. Develop concepts and skills in "mind-sized bites." Ensure that students understand the expectations. Including all students when implementing standards leads to increased emphasis on conducting experiments, authentic problem solving, and project-based learning (McLaughlin et al. 1999).

3. Build on children's strengths. Every child has his or her own strengths as a learner, such as visual–perceptual ability, communication skills, cooperative learning, or

organization. Teachers need to identify and then encourage the development of learning based on student strengths.

4. Geometric proofs challenge both our spatial abilities and our sequential-ordering abilities. Students may have difficulties with either aspect. Scaffolding, in which the teacher helps organize and decompose the problem into smaller "chunks," may help students having problems with sequential ordering. Repeating completed steps and chunks and making written or graphic records of these might also be helpful. Computer programs such as the Geometry Tutor furnish visual and other types of online help for just these tasks (Anderson et al. 1995; Schofield, Eurich-Fulcer, and Britt 1994). Perhaps most important, teachers must observe and talk to students to see which, if any, of these difficulties are holding back progress. Do students have difficulty interpreting the geometric figure? Finding embedded shapes? Or do they need assistance understanding the sequence of the arguments in a proof or how achieving each part of a proof related to other parts? Do they have difficulty remembering relevant postulates and theorems?

5. Use real objects, manipulatives, pictures, and diagrams wisely (Kapperman, Heinze, and Sticken 1997). They can help students with learning disabilities learn both concepts and skills (Mastropieri, Scruggs, and Shiah 1991). However, students (and teachers) can use even real objects and manipulatives in a rote manner if the emphasis is not on meaningful situations and tasks (Clements and McMillen 1996). Make sure that students explain what they are doing and link their work with manipulatives to all their other work in mathematics. Make connections and integrate concepts and skills. Use every possible social context to supply meaningful situations for mathematical problem solving (Parmar and Cawley 1997). Help children link symbols, verbal descriptions, work with ma-nipulatives, and everyday situations. Help them link number and arithmetic ideas to geometry, for example, counting the sides of polygons or measuring for maps. Computers can help make such connections, leading to the next recommenda-tion. An early and continued emphasis on structure and pattern in mathematics focused especially on geometric models and improving students' visual memory positively affects children at risk for later school failure (Fox 2006).

6. Use technology wisely. Computers can serve many purposes (Clements and Nastasi 1992; Mastropieri, Scruggs, and Shiah 1991; Pagliaro 1998; Shaw and Durden 1998; Swanson and Hoskyn 1998). As mentioned, computers can also serve as a valuable extension to traditional manipulatives that might be particu-larly helpful to students with special needs (Hutinger and Johanson 2000; Weir 1987). Using dynamic geometry software aided one high school student with cerebral palsy in understanding angles (Shaw and Durden 1998). Children who were blind and partially sighted used a computer-guided floor turtle to develop spatial concepts such as *right* and *left* (Gay 1989). Also, high-quality computer software can have advantages such as being patient and nonjudgmental, giving

undivided attention, proceeding at the child's pace, and offering immediate rein-forcement (Sarama and Clements 2002; Schery and O'Connor 1997). If software helps afford practice on an initially small set of items, such as shape names, and then expands as students master each shape, positive effects on learning for students with special needs are strong (Woodward 1995).

7. With and without technology, give judicious review that is sufficient, distributed, cumulative, and varied (Kameenui and Carnine 1998).

8. Plan with colleagues to create a coherent and connected special mathematics program (Pagliaro 1998; Parmar and Cawley 1997). Standards can help create a common language between special- and general education teachers (McLaughlin et al. 1999). For students with learning disabilities, we need to continually assess what the most important concepts and skills are—and look across the grades for their development. Consider emphasis on geometry and measurement (Parmar and Cawley 1997). Focus on the important ideas (Kameenui and Carnine 1998; NCTM 2006). All students need access to varied topics in mathematics. Topics beyond number and operations are increasingly important in our daily lives. Students with special needs usually do better in regular classrooms, but only if they receive differentiated instruction there (Gelzheiser et al. 2000).

9. Expand time for mathematics, including geometry. In general, a state or school district's curriculum does not allow adequate time for the many instructional and learning strategies necessary for the mathematical success of students with learning disabilities (Lerner 1997). Students with special needs need *additional* geometric and spatial experiences.

Areas of Focus and Instructional Activities

Geometry affords students a platform on which to learn and practice problem-solving skills with abstract and real-life applications. Instruction in geometry should enable students to identify and create geometric shapes, beginning with recognizing and sorting two- and three-dimensional shapes and continuing through using trigonometric functions to determine angle measurements. Children also learn to discuss properties and attributes of shapes, including sides, angles, area, and perimeter, and to make decisions concerning the relationships among these properties. This section covers the following areas of focus: (1) shape properties and angle measure; (2) location and position; and (3) shape composition, transformations, and spatial reasoning.

Teaching about shape properties and angle measures

The learning of a fifth-grade boy with learning disabilities illustrates the benefit of a curriculum that includes geometry (from Thornton, Langrall, and Jones 1997). Terrell was in a general fifth-grade class. He did not appear capable of abstract reasoning and had visual perception problems. He did seem to be able to retrieve information once he had learned it well.

The task was designed to develop students' understandings of angle measure. Angles are the turning points in the study of geometry and spatial relationships. Without a firm understanding of angles and angle measure, much of geometry is difficult or incomprehensible. Therefore, understanding angle measurement in degrees was on Terrell's individualized education plan. In this task, students were supposed to figure out the angle measures of the red trapezoid in the popular pattern block manipulative.

Terrell worked with a group that solved the problem of figuring out the measure of the large angle of the pattern block trapezoid. He arranged three trapezoids around a point on an overhead projector and showed it to the class (fig. 7.3). He explained that another student, Duane, had told their group about how a "360 flip" off the high diving board "goes all the way around." Terrell gestured to his illustration and argued that, because three trapezoids also "go all the way around," they could determine the angle of each by dividing 360 by 3, yielding 120° for the "big angle" of the trapezoid.

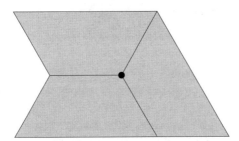

Fig. 7.3. Terrell's problem: three trapezoids arranged around a point

When the teacher was observing the students working in groups, he was concerned that Terrell might not have understood the notion of the 360 flip and its implications for the size of the trapezoids' angles. So, he was pleased that Terrell could explain the concept to the class in his own words. The teacher incorporated Terrell's group's solution in a summary chart that also included different solutions that other groups had constructed (Thornton, Langrall, and Jones 1997).

We see that despite his learning disabilities, Terrell used *manipulatives meaningfully* and used *connections* to real-world situations, here a flip off a diving board, to reason abstractly about angle measures in a way that was meaningful to him. The vignette also illustrates that the teacher held *high expectations* for students' learning of geometry and used *peer interaction*, *problem solving*, and communication, which support learning and memory, as worthwhile pedagogical strategies. Similarly, had a student with significant cognitive disabilities been involved in this lesson, that student could have contributed by counting and recording the number of pattern blocks that the group used.

Of course, the teacher should not leave the topic after only a few experiences. *Follow-up tasks and practice* could include determining the angles of other pattern block and tangram shapes. As part of that work, *computer software* that allows children to rotate shapes in predetermined increments (e.g., 15°) and manipulate and measure the angles of shapes (fig. 7.4) can be particularly helpful (Sarama, Clements, and Vukelic 1996).

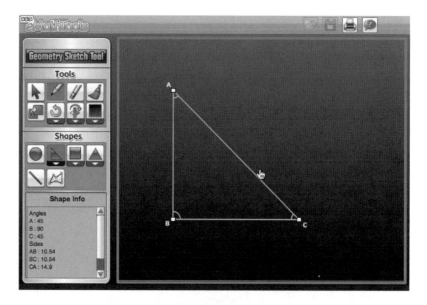

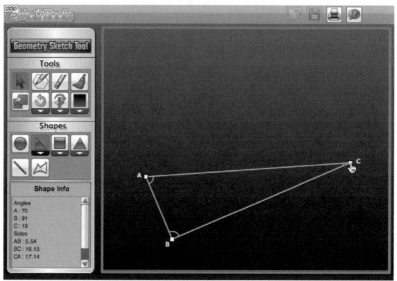

Fig. 7.4. Two screens from the Building Blocks software (eTools suite, SRA/ McGraw–Hill), allowing children to define, manipulate, turn, and measure geometric figures. When the triangle defined as a right triangle is manipulated (a good spatial activity in itself), the measures update automatically.

Teaching about position and location

The setting for this lesson on position and location is a sixth-grade mathematics class with sixteen students, nine of whom are classified as having learning disabilities, with various specific learning difficulties. Students are giving directions from one location to another on a map of the local area. The activity involves five steps: (1) identify the two locations marked on the map and indicate their relative geographic location to each other (e.g., east–

southeast; due north); (2) with a highlighter, trace the shortest path from one location to the other; (3) with another color highlighter, trace the path that maximizes the use of a highway or major road; (4) write out directions for a car driving from one location to another; and (5) pretend to call a group member on the phone and accurately relay the directions while the other person follows along on another map.

Within the goal of developing students' sense of spatial relations in a real-world application, the teacher addressed the learning needs of individual students. For example, Sue, a student with memory difficulties, had the role of writing out the directions. The teacher asked Janet, a student with difficulty maintaining attention, to give the directions on the phone. Joe, a student with difficulties with communication, led a discussion on how the spatial relation of point A to point B is the compass opposite of the relation of point B to point A. The activity used manipulatives (map, diagram of compass) and learning aids (listing of directions, diagram with a mnemonic for *left* and *right*, highlighters of different colors) and encouraged cooperative learning, with teacher guidance only when necessary. Thus, the teacher focused on geometric and spatial skills, as well as the *Principles and Standards* processes of reasoning and communication abilities within the topic area.

Teaching about shape composition, transformations, and spatial reasoning

The ability to describe, use, and visualize the effects of composing and decomposing shapes is a major competence. It is relevant to many topics not only in geometry but also in the sciences and arts. Further, the concepts and actions of creating and then iterating units and higher-order units when constructing patterns, measuring, and computing are established bases for mathematical understanding and analysis. This type of composition corresponds with, and supports, children's ability to compose and decompose numbers.

A class of students with severe learning disabilities had this competence on their individualized education plans (from Thornton, Langrall, and Jones 1997). The teacher planned an activity around the following problem: "Is every triangle half of a rectangle? Prove it." The teacher supplied two copies of three triangular regions (fig. 7.5), one set white, one colored.

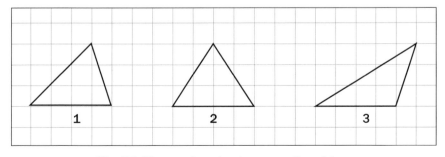

Fig. 7.5. Three triangles in a proof problem

To begin, three boys worked together to cut out the colored triangles. They then taped them to the sheet with the white triangles and formed parallelograms. Their premise was "No—two equal triangles do not form rectangles. The shapes formed are not rectangles because they do not have 90-degree angles" (from Stone 1993, p. 54, as reported in Thornton, Langrall, and Jones 1997).

A different pair of students cut the triangles on their altitudes and taped the two colored triangles, one on both sides, to the white triangle. They had a difficulty with the third, obtuse triangle, so they cut off a section. After they taped the two pieces to the existing triangle they had a small piece sticking out on the left and a small hole on the right (fig. 7.6). They asked whether they could cut off the piece and move it (fig. 7.7). Given permission, they produced a rectangle.

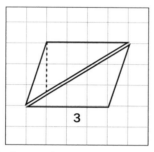

Fig. 7.6. A step in students' solving the triangle–rectangle proof problem

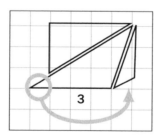

Fig. 7.7. Students cut off and move part of one triangle in solving the triangle–rectangle proof problem.

Finally, a girl, working independently (because someone was absent), also cut the triangles. Her first idea was that all triangles except #3 could form rectangles. She was proud when she finally figured out how to do #3 like the boys did.

Thus, although they used different assumptions, the two different solutions were both valid. If one assumes that the triangles could not be decomposed, then it was not possible for every triangle to be half of a rectangle. If one could decompose, then it is possible.

In summary, these students with severe learning disabilities engaged in mathematical *reasoning* as they solved a *meaningful problem* addressing an important area of focus in geometry. The teacher's expectations were *reasonable, but not low*. Students could build on the *strengths* they had to solve the problem in their own ways.

Again, this was the beginning, not the end, of work on this topic. Recall that the first group of students eventually hypothesized that every triangle is half a parallelogram but not necessary half a rectangle. In a *follow-up* lesson, the teacher had students try to prove or disprove this new question. The second group similarly tried to see whether they could form a rectangle by decomposing two congruent triangles. The students thereby *practiced* what they had learned and extended it to other areas of focus in geometry, such as the properties of geometric shapes, in struggling with the definitions of parallelogram and rectangle, as well as in the distinction between congruent shapes and shapes that are not congruent but have the same area. In later activities, the teacher could make *connections* between these activities and ideas and the areas of triangles and rectangles. That is, she could help the students understand and apply the relationship between the length and width measures of a rectangle and the corresponding base and height measures of a triangle (Thornton, Langrall, and Jones 1997). She might use a computer software program to decompose, recompose, and measure these figures.

Many other ways exist for teachers to adapt geometry activities in teaching children with special needs. Table 7.1 shows a few general suggestions. Finally, figure 7.8 shows a sample from a suite of activities that follow research-based learning trajectories.

Table 7.1
Specific learning difficulties related to a given geometry activity

Difficulty with:	Potential behavior indicators	Teaching suggestions
Attending to tasks	Playing with manipulatives as if they were toys rather than engaging in the mathematical task; looking around the room	Have the student take an active role in construction; quicken pace of lesson for shorter downtime; give only as many materials as necessary; have student self-monitor time on task; break task into smaller steps to allow for a feeling of completion; furnish only necessary manipulatives in smaller clusters
Keeping place on a page in the text or workbook	Losing directions; not knowing where to read; dropping books and pencils in frustration	Give directions for each activity on a separate page; highlight parts most directly relevant to the student with disabilities; assign a "buddy" to help with getting started

Table 7.1—*Continued*

Difficulty with:	Potential behavior indicators	Teaching suggestions
Using manipulatives or construction tools	Inaccurate constructions; frequent building and rebuilding, indicating frustration; throwing objects	Modify tools, such as using larger shapes; review use of manipulatives or tools with entire class before lesson; break task into smaller units or steps
Following a sequence of steps to solution	Errors in solutions and solution processes; guessing at answers; messy paper with many erasures	Supply a reference card with procedures for student to emulate; use mnemonics where appropriate to assist memory; furnish manipulative materials
Oral or written communication in mathematics	Using "everyday" rather than mathematical vocabulary; avoiding writing or speaking activities in the classroom	Break down writing or speaking requirements into segments; give frequent opportunities for communication; reinforce use of mathematical vocabulary; post vocabulary and definitions, using pictures when possible; encourage alternate representations using the computer or diagrams
Reading	Frustration with reading; inability to read written problems	Teach sight word vocabulary relevant to geometry; preteach vocabulary relevant to geometric concept; review key vocabulary before lesson
Determining relevant information	Distracted by irrelevant information in written problems	Teach highlighting technique; demonstrate strategies using "think aloud"
Interpreting pictures and diagrams	Incorrect labeling of information on a diagram; inability to draw a representation of a figure	Have student verbalize information on diagram; furnish an illustrative model for drawing
Memorizing facts and formulas	Frustration with arithmetic and problems involving formulas	When using formulas, allow students to use a calculator to compute at each step in the formula; focus on arithmetic facts at other times; give a mnemonic or memory aid

Geometry Snapshots 3. Students identify an image that matches the "symmetric whole" of a target image from four multiple-choice selections. In other levels, students match shapes or combinations of shapes that have been turned or flipped.

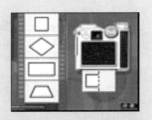

Mystery Pictures 4. Children build pictures by identifying shapes that are named by the Building Blocks software program. (Mystery Pictures 3 is appropriate before this activity; it teaches the shape names while asking children to match shapes in different orientations.)

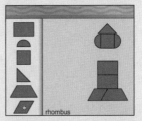

Shape Parts 1 and 7. Students use shape parts to construct a shape that matches a target shape (level 1) or a description (level 7).

Shape Shop 1. Students identify shapes by their attributes or number of parts (e.g., number of sides and angles).

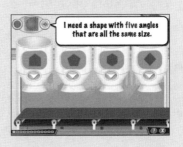

Fig. 7.8. Suite of computer activities from the Building Blocks software (Clements and Sarama 2007)—*Continues*

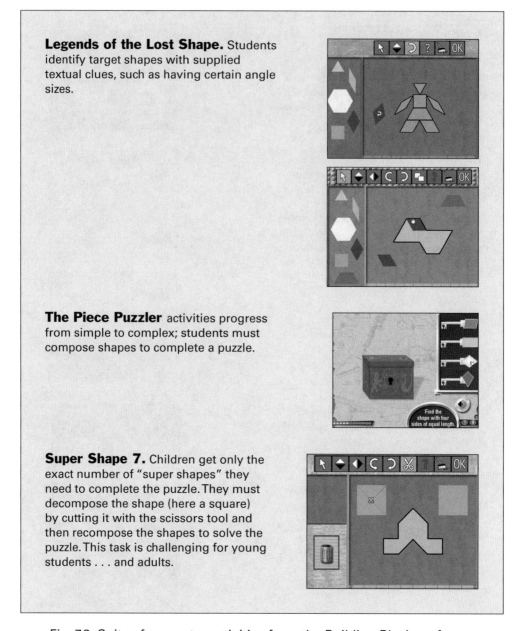

Legends of the Lost Shape. Students identify target shapes with supplied textual clues, such as having certain angle sizes.

The Piece Puzzler activities progress from simple to complex; students must compose shapes to complete a puzzle.

Super Shape 7. Children get only the exact number of "super shapes" they need to complete the puzzle. They must decompose the shape (here a square) by cutting it with the scissors tool and then recompose the shapes to solve the puzzle. This task is challenging for young students . . . and adults.

Fig. 7.8. Suite of computer activities from the Building Blocks software (Clements and Sarama 2007)—*Continued*

Final Words

NCTM's mission is to ensure equitable mathematics learning of the highest quality for all students, regardless of their personal characteristics, backgrounds, or physical challenges. All students must have opportunities to study—and support to learn—mathematics. *Equitable* does not mean that every student should receive identical instruction; instead, it demands reasonable and appropriate accommodations as needed to promote access and attainment for all students (NCTM 2000, p. 12). We are not yet succeeding on this goal. "Children with disabilities remain those most at risk of being left behind" (U.S. Department of Education 2002, p. 4). We believe that the geometric and spatial domains offer a means to allow children with special needs to engage in substantive mathematical reasoning, problem solving, and learning.

Reflections on Teaching Geometry: Questions for Discussion

1. This chapter opened by introducing the issue that the educational programming for students with disabilities has privileged instruction focused on number and operations over geometry. Do you believe that such a prioritization should exist? What are some points where one area could complement the other in students' learning experiences?

2. In what ways are geometry tasks challenging specifically for students with *different* disabilities?

3. What general supports can benefit students with disabilities learning geometric and spatial concepts and skills? What specific supports would you offer students with specific disabilities?

4. Why should teachers of students with disabilities use strategies other than simply slowing the pace of instruction and supplying manipulatives? What do you see as the difference between a manipulative used for instruction and one used to overcome the effects of a particular disability?

References

Anderson, John R., Albert T. Corbett, Kenneth R. Koedinger, and Ray Pelletier. "Cognitive Tutors: Lessons Learned." *Journal of the Learning Sciences* 4, no. 2 (1995): 167–207.

Arditi, Aries, Jeffrey D. Holtzman, and Stephen M. Kosslyn. "Mental Imagery and Sensory Experience in Congenital Blindness." *Neuropsychologia* 26, no. 1 (1988): 1–12.

Battista, Michael T., and Douglas H. Clements. "Students' Understanding of Three-Dimensional Rectangular Arrays of Cubes." *Journal for Research in Mathematics Education* 27 (May 1996): 258–92.

Battista, Michael T., Douglas H. Clements, Judy Arnoff, Kathryn Battista, and Caroline Van Auken Borrow. "Students' Spatial Structuring of 2D Arrays of Squares." *Journal for Research in Mathematics Education* 29 (November 1998): 503–32.

Bellugi, Ursula, Liz Lichtenberger, Wendy Jones, Zona Lai, and Marie St. George. "The Neurocognitive Profile and Williams Syndrome: A Complex Pattern of Strengths and Weaknesses." *Journal of Cognitive Neuroscience* 12 (March 2000, Suppl. 1): 7–29.

Burger, William F., and J. Michael Shaughnessy. "Characterizing the van Hiele Levels of Development in Geometry." *Journal for Research in Mathematics Education* 17 (January 1986): 31–48.

Clements, Douglas H. "Training Effects on the Development and Generalization of Piagetian Logical Operations and Knowledge of Number." *Journal of Educational Psychology* 76 (October 1984): 766–76.

Clements, Douglas H., and Sue McMillen. "Rethinking 'Concrete' Manipulatives." *Teaching Children Mathematics* 2 (January 1996): 270–79.

Clements, Douglas H., and Bonnie K. Nastasi. "Computers and Early Childhood Education." In *Preschool and Early Childhood Treatment Directions*, edited by Maribeth Gettinger, Stephen N. Elliott, and Thomas R. Kratochwill, pp. 187–246. Hillsdale, N.J.: Lawrence Erlbaum, 1992.

Clements, Douglas H., and Julie Sarama. Building Blocks [computer software]. Columbus, Ohio: SRA/McGraw–Hill, 2007.

Clements, Douglas H., Julie Sarama, and Ann-Marie DiBiase, eds. *Engaging Young Children in Mathematics: Standards for Early Childhood Mathematics Education*. Mahwah, N.J.: Lawrence Erlbaum, 2004.

Clements, Douglas H., Sudha Swaminathan, Mary Anne Zeitler Hannibal, and Julie Sarama. "Young Children's Concepts of Shape." *Journal for Research in Mathematics Education* 30 (March 1999): 192–212.

Council of Chief State School Officers (CCSSO). *Common Core State Standards for Mathematics.* Washington, D.C.: CCSSO and the National Governors Association Center for Best Practices, 2010. www.corestandards.org.

Dixon, Juli Kim. "Limited English Proficiency and Spatial Visualization in Middle School Students' Construction of the Concepts of Reflection and Rotation." *Bilingual Research Journal* 19 (Spring 1995): 221–47.

Driscoll, Mark J. *Research within Reach: Elementary School Mathematics and Reading.* St. Louis: CEMREL, 1983.

Fox, J. "A Justification for Mathematical Modelling Experiences in the Preparatory Classroom." In *Proceedings of the 29th Annual Conference of the Mathematics Education Research Group of Australia*, edited by Peter Grootenboer, Robyn Zevenbergen, and Mohan Chinnappan, pp. 221–28. Canberra, Australia: MERGA, 2006.

Freudenthal, Hans. *Mathematics as an Educational Task.* Dordrecht, Netherlands: Reidel, 1973.

Fuys, David, Dorothy Geddes, and Rosamond Tischler. "The van Hiele Model of Thinking in Geometry among Adolescents." *Journal for Research in Mathematics Education Monograph Number 3.* Reston, Va.: National Council of Teachers of Mathematics, 1988.

Gay, Penny. "Tactile Turtle: Explorations in Space with Visually Impaired Children and a Floor Turtle." *British Journal of Visual Impairment* 7 (Spring 1989): 23–25.

Geary, David C., Mary K. Hoard, and Carmen O. Hamson. "Numerical and Arithmetical Cognition: Patterns of Functions and Deficits in Children at Risk for a Mathematical Disability." *Journal of Experimental Child Psychology* 74 (November 1999): 213–39.

Gelzheiser, Lynn M., Bonnie A. Griesemer, Robert M. Pruzek, and Joel Meyers. "How Are Developmentally Appropriate or Traditional Teaching Practices Related to the Mathematics Achievement of General and Special Education Students?" *Early Education and Development* 11 (March 2000): 217–38.

Gibson, Eleanor J. *Principles of Perceptual Learning and Development.* New York: Appleton-Century-Crofts, Meredith Corp., 1969.

Gravemeijer, Koeno P. E. "Realistic Geometry Instruction." In *Contexts Free Productions: Tests and Geometry in Realistic Mathematics Education*, edited by Koeno P. E. Gravemeijer, Marja van den Heuvel, and Leen Streefland, pp. 79–91. Utrecht, Netherlands: OW & OC, 1990.

Greabell, Leon C. "The Effect of Stimuli Input on the Acquisition of Introductory Geometric Concepts by Elementary School Children." *School Science and Mathematics* 78 (April 1978): 320–26.

Grobecker, Betsey. "Mathematics Reform and Learning Differences." *Learning Disability Quarterly* 22 (Winter 1999): 43–58.

Gross-Tsur, Varda, Ruth S. Shalev, Orly Manor, and Naomi Amir. "Developmental Right-Hemisphere Syndrome: Clinical Spectrum of the Nonverbal Learning Disability." *Journal of Learning Disabilities* 28 (February 1995): 80–86.

Howlin, Patricia, Mark Davies, and Oriee Udwin. "Syndrome Specific Characteristics in Williams Syndrome: To What Extent Do Early Behavioural Patterns Persist into Adult Life?" *Journal of Applied Research in Intellectual Disabilities* 11 (September 1998): 207–26.

Hutinger, Patricia L., and Joyce Johanson. "Implementing and Maintaining an Effective Early Childhood Comprehensive Technology System." *Topics in Early Childhood Special Education* 20 (Fall 2000): 159–73.

Iverson, Jana M., and Susan Goldin-Meadow. "What's Communication Got to Do with It? Gesture in Children Blind from Birth." *Developmental Psychology* 33 (May 1997): 453–67.

Kameenui, Edward J., and Douglas W. Carnine. *Effective Teaching Strategies That Accommodate Diverse Learners*. Upper Saddle River, N.J.: Prentice-Hall, 1998.

Kamii, Constance K., Judith Rummelsburg, and Amy Kari. *Teaching Arithmetic to Low-Performing, Low-SES First Graders*. San Diego: American Educational Research Association, 2004.

Kapperman, Gaylen, Toni Heinze, and Jodi Sticken. *Strategies for Developing Mathematics Skills in Students Who Use Braille*. Sycamore, Ill.: Research and Development Institute, 1997.

Keller, Shirley, and Irma Goldberg. *Let's Learn Shapes with Shapely-CAL*. Great Neck, N.Y.: Creative Adaptations for Learning, 1997.

Landau, Barbara. "The Construction and Use of Spatial Knowledge in Blind and Sighted Children." In *Spatial Cognition: Brain Bases and Development*, edited by Joan Stiles-Davis, Mark Kritchevsky, and Ursula Bellugi, pp. 343–71. Mahwah, N.J.: Lawrence Erlbaum, 1988.

Landau, Barbara, H. Gleitman, and E. Spelke. "Spatial Knowledge and Geometric Representation in a Child Blind from Birth." *Science* 213 (September 1981): 1275–77.

Lebron-Rodriguez, Delia Ester, and Robert Pasnak. "Induction of Intellectual Gains in Blind Children." *Journal of Experimental Child Psychology* 24 (December 1977): 505–15.

Lerner, Janet. *Learning Disabilities*. Boston: Houghton Mifflin, 1997.

Mason, Marguerite M. "Geometric Knowledge in a Deaf Classroom: An Exploratory Study." *Focus on Learning Problems in Mathematics* 17 (Summer 1995): 57–69.

Mastropieri, Margo A., Thomas E. Scruggs, and S. Shiah. "Mathematics Instruction for Learning-Disabled Students: A Review of Research." *Learning Disabilities Research and Practice* 6, no. 2 (1991): 89–98.

McLaughlin, Margaret J., Victor Nolet, Lauren Morando Rhim, and Kelly Henderson. "Integrating Standards Including All Students." *Teaching Exceptional Children* 31 (January–February 1999): 66–71.

McLean, Janet F., and Graham J. Hitch. "Working Memory Impairments in Children with Specific Arithmetic Learning Difficulties." *Journal of Experimental Child Psychology* 74 (November 1999): 240–60.

McMurray, Nancy E., Michael E. Bernard, Herbert J. Klausmeier, J. M. Schilling, and Katherine Vorwerk. "Instructional Design for Accelerating Children's Concept Learning." *Journal of Educational Psychology* 69 (December 1977): 660–67.

Millar, Susanna, and Miriam Ittyerah. "Movement Imagery in Young and Congenitally Blind Children: Mental Practice without Visuo-Spatial Information." *International Journal of Behavioral Development* 15 (March 1992): 125–46.

Mix, Kelly S., Janellen Huttenlocher, and Susan Cohen Levine. *Quantitative Development in Infancy and Early Childhood*. New York: Oxford University Press, 2002.

Morrongiello, Barbara A., Brian Timney, G. Keith Humphrey, Suzanne Anderson, and Cheryl Skory. "Spatial Knowledge in Blind and Sighted Children." *Journal of Experimental Child Psychology* 59 (April 1995): 211–33.

Mullet, Etienne, and Robert Miroux. "Judgment of Rectangular Areas in Children Blind from Birth." *Cognitive Development* 11 (January–March 1996): 123–39.

National Council of Teachers of Mathematics (NCTM). *Principles and Standards for School Mathematics*. Reston, Va.: NCTM, 2000.

———. *Curriculum Focal Points for Prekindergarten through Grade 8 Mathematics: A Quest for Coherence*. Reston, Va.: NCTM, 2006.

National Mathematics Advisory Panel. *Foundations for Success: The Final Report of the National Mathematics Advisory Panel*. Washington D.C.: U.S. Department of Education, Office of Planning, Evaluation, and Policy Development, 2008.

Newcombe, Nora S., and Janellen Huttenlocher. *Making Space: The Development of Spatial Representation and Reasoning*. Cambridge, Mass.: MIT Press, 2000.

Olson, J. K. "Microcomputers Make Manipulatives Meaningful." Paper presented at the meeting of the International Congress of Mathematics Education, Budapest, Hungary, August 1988.

Pagliaro, Claudia M. "Mathematics Reform in the Education of Deaf and Hard of Hearing Students." *American Annals of the Deaf* 143 (March 1998): 22–28.

Parmar, René S., and John F. Cawley. "Preparing Teachers to Teach Mathematics to Students with Learning Disabilities." *Journal of Learning Disabilities* 30 (March–April 1997): 188–197.

Pasnak, Robert, S. E. Madden, Valerie A. Malabonga, and R. Holt. "Persistence of Gains from Instruction in Classification, Seriation, and Conservation." *Journal of Educational Research* 90 (November–December 1996): 87–92.

Peterson, J. Michael, and Mishael M. Hittie. *Inclusive Teaching: Creating Effective Schools for All Learners*. Boston: Allyn and Bacon, 2003.

Raphael, Dennis, and Merlin Wahlstrom. "The Influence of Instructional Aids on Mathematics Achievement." *Journal for Research in Mathematics Education* 20 (March 1989): 173–90.

Rourke, Byron P. "Studies of Persons with Learning Disabilities at the University at Windsor Laboratory: The First 25 Years." *National* 28 (1991): 41–44.

Rourke, Byron P., and James A. Conway. "Disabilities of Arithmetic and Mathematical Reasoning: Perspectives from Neurology and Neuropsychology." *Journal of Learning Disabilities* 30 (January–February 1997): 34–46.

Rourke, Byron P., and M. Alan J. Finlayson. "Neuropsychological Significance of Variations in Patterns of Academic Performance: Verbal and Visual-Spatial Abilities." *Journal of Abnormal Child Psychology* 6 (March 1978): 121–33.

Sarama, Julie, and Douglas H. Clements. "Learning and Teaching with Computers in Early Childhood Education." In *Contemporary Perspectives in Early Childhood Education*, edited by Olivia N. Saracho and Bernard Spodek, pp. 171–219. Greenwich, Conn.: Information Age Publishing, 2002.

Sarama, Julie, Douglas H. Clements, Sudha Swaminathan, Sue McMillen, and Rosa M. González Gómez. "Development of Mathematical Concepts of Two-Dimensional Space in Grid Environments: An Exploratory Study." *Cognition and Instruction* 21, no. 3 (2003): 285–324.

Sarama, Julie, Douglas H. Clements, and Elaine Bruno Vukelic. "The Role of a Computer Manipulative in Fostering Specific Psychological/Mathematical Processes." In *Proceedings of the Eighteenth Annual Meeting of the North America Chapter of the International Group for the Psychology of Mathematics Education*, edited by Elizabeth Jakubowski, Dierdre Watkins, and Harry Biske, vol. 2, pp. 567–72. Columbus, Ohio: ERIC Clearinghouse for Science, Mathematics, and Environmental Education, 1996.

Schery, Teris K., and Lisa C. O'Connor. "Language Intervention: Computer Training for Young Children with Special Needs." *British Journal of Educational Technology* 28 (October 1997): 271–79.

Schofield, Janet Ward, Rebecca Eurich-Fulcer, and Chen L. Britt. "Teachers, Computer Tutors, and Teaching: The Artificially Intelligent Tutor as an Agent for Classroom Change." *American Educational Research Journal* 31 (January 1994): 579–607.

Semrud-Clikeman, Margaret, and George W. Hynd. "Right Hemispheric Dysfunction in Nonverbal Learning Disabilities: Social, Academic, and Adaptive Functioning in Adults and Children." *Psychological Bulletin* 107 (March 1990): 196–209.

Share, David L., Terrie E. Moffitt, and Phil A. Silva. "Factors Associated with Arithmetic-and-Reading Disabilities and Specific Arithmetic Disability." *Journal of Learning Disabilities* 21 (May 1988): 313–20.

Shaw, Kenneth L., and Paul Durden. "Learning How Amanda, a High School Cerebral Palsy Student, Understands Angles." *School Science and Mathematics* 98 (April 1998): 198–204.

Shaw, R. A. "Curriculum and Instructional Activities: The Upper Grades." In *Developmental Teaching of Mathematics for the Learning Disabled*, edited by John F. Cawley, pp. 145–205. Rockville, Md.: Aspen Systems Corp., 1984.

Sliva, Julie A. *Teaching Inclusive Mathematics to Special Learners, K–6*. Thousand Oaks, Calif.: Corwin Press, 2004.

Sousa, David A. *How the Special Needs Brain Learns*. Thousand Oaks, Calif.: Corwin, 2001.

Sowell, Evelyn J. "Effects of Manipulative Materials in Mathematics Instruction." *Journal for Research in Mathematics Education* 20 (November 1989): 498–505.

Spiers, P. A. "Alcalculia Revisited: Current Issues." In *Mathematical Disabilities: A Cognitive Neuropsychological Perspective*, edited by Gerard Deloche and Xavier Seron. Hillsdale, N.J.: Lawrence Erlbaum, 1987.

Stewart, R., N. Leeson, and R. J. Wright. "Links between Early Arithmetical Knowledge and Early Space and Measurement Knowledge: An Exploratory Study." In *Proceedings of the Twentieth Annual Conference of the Mathematics Education Research Group of Australasia*, edited by F. Biddulph and K. Carr, vol. 2, pp. 477–84. Hamilton, New Zealand: MERGA, 1997.

Suydam, Marilyn N. "Research Report: Manipulative Materials and Achievement." *Arithmetic Teacher* 33 (February 1986): 10–32.

Swanson, H. Lee. "An Information Processing Analysis of Learning Disabled Children's Problem Solving." *American Educational Research Journal* 30 (Winter 1993): 861–93.

Swanson, H. Lee, and Maureen Hoskyn. "Experimental Intervention Research on Students with Learning Disabilities: A Meta-Analysis of Treatment Outcomes." *Review of Educational Research* 68 (Fall 1998): 277–321.

Thornton, Carol A., Cynthia W. Langrall, and Graham A. Jones. "Mathematics Instruction for Elementary Students with Learning Disabilities." *Journal of Learning Disabilities* 30 (March–April 1997): 142–50.

Tranel, Daniel, Laurie E. Hall, Steve Olson, and Ned N. Tranel. "Evidence for a Right-Hemisphere Developmental Learning Disability." *Developmental Neuropsychology* 3, no. 2 (1987): 113–27.

U.S. Department of Education. Office of Special Education and Rehabilitative Services. "A New Era: Revitalizing Special Education for Children and Their Families." 2002. www.ed.gov/inits /commissionsboards/whspecialeducation (accessed May 3, 2010).

van Hiele, Pierre M. *Structure and Insight: A Theory of Mathematics Education*. Orlando, Fla.: Academic Press, 1986.

van Hiele-Geldof, Dina. "The Didactics of Geometry in the Lowest Class of Secondary School" (M. Verdonck, trans.). In *English Translation of Selected Writings of Dina van Hiele-Geldof and Pierre M. van Hiele*, edited by David Fuys, Dorothy Geddes, and Rosamond Tischler, pp. 1–214. Brooklyn, N.Y.: Brooklyn College, School of Education, 1984. ERIC Document Reproduction no. ED287697.

Wansart, William L. "Learning to Solve a Problem: A Microanalysis of the Solution Strategies of Children with Learning Disabilities." *Journal of Learning Disabilities* 23 (March 1990): 164–70, 184.

Weir, Sylvia. *Cultivating Minds: A Logo Casebook.* New York: Harper and Row, 1987.

Wenglinsky, Harold. *How Teaching Matters: Bringing the Classroom Back into Discussions of Teacher Quality*: Princeton, N.J.: Educational Testing Service, 2000.

Woodward, John. "Technology-Based Research in Mathematics for Special Education." *Focus on Learning Problems in Mathematics* 17 (Spring 1995): 3–23.

CHAPTER 8

Measurement

René S. Parmar,
René Garrison,
Douglas H. Clements,
and Julie Sarama

As one of the primary uses of mathematics in everyday life, measurement is an essential topic for all students. This chapter addresses how students with disabilities can access the concepts and tools of measurement. We give an overview of the National Council of Teachers of Mathematics (NCTM) *Principles and Standards for School Mathematics* (NCTM 2000) for measurement and *Curriculum Focal Points for Prekindergarten through Grade 8 Mathematics: A Quest for Coherence* (NCTM 2006), discuss the difficulties in measurement that students with disabilities encounter, and offer examples of instructional techniques and modifications to meet the needs of students with disabilities. Although we developed the chapter before the release of the *Common Core State Standards* (Council of Chief School State Officers [CCSSO] 2010), the discussion and examples of instructional techniques and modifications parallel the standards in the Common Core.

The *Principles and Standards* key aspects of measurement (NCTM 2000, pp. 44–47) are as follows:

1. Understand measurable attributes of objects and the units, systems, and processes of measurement. In the early grades, this aspect includes recognizing the attributes of length, volume, weight, area, and time in addition to comparing and ordering objects according to a given attribute. With an emphasis on length, young children learn how to measure using nonstandard and standard units and to select an appropriate unit and tool for the attribute being measured. In intermediate grades, students should develop familiarity with standard units in the customary and metric systems appropriate for measuring different attributes; carry out simple unit conversions within a system, such as from centimeters to meters; understand that measurements are approximations; understand how differences in units affect precision; and explore what happens to the measurements of two-dimensional shapes (perimeter and area) when shape changes in some way. Concepts extend to

the measurement of angles, perimeter, area, surface area, and volume. Also, middle school students explore derived measures, such as speed. High school students learn about scale and complex units such as pounds per square inch.

2. Apply appropriate techniques, tools, and formulas to determine measurements. These skills begin to develop in the early grades, with children learning to measure with multiple copies of units the same size, using repetition of a single unit to measure something larger than the unit, using tools to measure, and developing common referents for measures to compare and estimate. In the intermediate and middle school grades, students are expected to develop strategies for estimating the perimeters, areas, and volumes of irregular shapes; select and apply appropriate standard units and tools to measure length, area, volume, weight, time, temperature, and size of angles; select and use benchmarks to estimate measurements; develop, understand, and use formulas to find the area of rectangles and related triangles and parallelograms; and develop strategies to determine the surface areas and volumes of rectangular solids. In middle school the concepts and skills extend to selecting and applying techniques and tools to accurately find length, area, volume, and angle measures to appropriate levels of precision; developing and using formulas to determine the circumference of circles and the area of triangles, parallelograms, trapezoids, and circles; developing strategies to find the area of more complex shapes; developing strategies to determine the surface area and volume of selected prisms, pyramids, and cylinders; solving problems involving scale factors; using ratio and proportion; and solving simple problems involving rates and derived measurements for such attributes as velocity and density. High school students use formulas in solving problems, including using algebraic procedures to organize conversions and computations by using unit analysis.

The Importance of Measurement

Measurement is one of the most common applications of mathematics: combining the concepts of geometry and number (NCTM 2000, 2006). NCTM's *Curriculum Focal Points* has defined the vital concepts related to measurement (fig. 8.1), along with examples of integrating measurement with other areas in mathematics.

Prekindergarten
Identifying measurable attributes and comparing objects by using these attributes (e.g., identifying "the same" or "different," and then "more" or "less," attributes such as length and weight, and solving problems by making direct comparisons of objects).

Kindergarten
Ordering objects by measurable attributes (e.g., solving problems by comparing and ordering objects by length).

Fig. 8.1. Curriculum Focal Points for measurement, pre-K–8

Grade 1
Solving problems involving measurements and data (e.g., laying multiple copies of a unit end to end and then counting the units by using groups of tens and ones, representing measurements and discrete data in picture and bar graphs).

Grade 2
Developing an understanding of linear measurement and facility in measuring lengths (e.g., understanding such underlying concepts as partitioning and transitivity, linear measure as an iteration of units, and use of rulers and other measurement tools).

Grade 3
Developing facility in measuring with fractional parts of linear units (e.g., analyzing attributes and properties of two-dimensional objects, understanding of perimeter as a measurable attribute and selecting appropriate units, strategies, and tools to solve problems involving perimeter).

Grade 4
Developing an understanding of area and determining the areas of two-dimensional shapes (e.g., selecting appropriate units, strategies [e.g., decomposing shapes], and tools for solving problems that involve estimating or measuring area, connecting area measure to multiplication, and justifying the formula for the area of a rectangle).

Grade 5
Describing three-dimensional shapes and analyzing their properties, including volume and surface area (e.g., analyzing properties of polyhedral solids; recognizing volume as an attribute of three-dimensional space; selecting appropriate units, strategies, and tools for solving problems that involve estimating or measuring volume and surface area).

Grade 6
Solving problems that involve areas and volumes; applying new work with equations in context.

Grade 7
Developing an understanding of and using formulas to determine surface areas and volumes of three-dimensional shapes (e.g., decomposing two- and three-dimensional shapes into smaller, component shapes to find surface areas; develop and justify formulas for the surface areas and volumes of prisms and cylinders).

Grade 8
Analyzing two- and three-dimensional space and figures by using distance and angle (e.g., using fundamental facts about distance and angles to describe and analyze figures and situations in two- and three-dimensional space and to solve problems, including those with multiple steps, explaining why the Pythagorean theorem is valid by using a variety of methods).

Fig. 8.1. Curriculum Focal Points for measurement, pre-K–8—*Continued*

A strong foundation in measurement knowledge and skills is essential for students' development of more sophisticated aspects of measurement such as making comparisons, forming estimations, and using formulas. To function in the world, people must be able to measure time, distance, weight, volume, money, and mass. Furthermore, people often must estimate measurements, as when they approximate how long getting from their home to the doctor's office takes to arrive on time for an appointment. People must also compare measurements to perform such tasks as shopping for the best bargain or determining the shortest route to work.

Measurement is relevant to many areas of learning beyond mathematics. Measurement is a crucial part of scientific experimentation. Every branch of science uses measurement tools for research and development. In social studies, an understanding of time is essential, as are measurements pertinent to geographical features that affect historical events. In the arts, measurement is an integral part of the design and construction of artistic products.

In mathematics, measurement plays a role in developing strategies to represent and manipulate information about the environment (Forrester, Latham, and Shire 1990). Thus, measurement is a tool for understanding quantitative aspects of our world. Measurement concepts rooted in geometrical representations can support early learning of number and later learning of algebra and trigonometry. Further, the ability to construct reasonable estimates, which allows us to interact with the environment daily, is an outgrowth of a strong understanding of measurement.

From the standpoint of teaching and learning, measurement is an interesting topic, offering many opportunities for instruction that uses concrete materials and makes connections to the everyday lives of children. These aspects are particularly essential for students with learning disabilities, who often have difficulty with the more routine tasks of paper–pencil arithmetic because of problems with reading or writing, focusing attention on the task, remaining sustained on task, and memorizing rules or routines. Learning activities focused on measurement offer many opportunities for a high level of student engagement and for physical and visual representations of concepts, both of which are vital principles for teaching students at risk for school failure (Fuchs and Fuchs 2001).

Big Ideas: Units, Use of Tools, Comparisons

To effectively use measurement, students must develop facility with three major aspects.

1. Units: standard and nonstandard units, appropriate units for measurement, using repetitions of a unit, units in customary and metric systems, unit conversions, solving problems by using units

2. Tools: using tools to measure attributes, using formulas to determine attributes, using common benchmarks

3. Comparisons: common referents for comparisons and estimates, scale, ratio, proportion

Students with disabilities may require adapted materials, specialized instruction, or alternative objectives to develop competency in these three aspects of measurement.

Students with learning disabilities may encounter difficulty learning names and characteristics of different units of measurement and in memorizing formulas. Some students with auditory processing difficulties will have trouble learning these concepts in a teacher-led lecture format. Students with attention deficit–hyperactivity disorder or attention deficit disorder may experience difficulty learning several terms or concepts presented in the same lesson.

The following dialog illustrates a second-grade student with disabilities who understands the underlying principles of various measurement scales but cannot express this understanding. The student does not use the appropriate comparative words (*longer, heavier, larger*), nor does he refer to using measurement tools for making a determination. Instead, he is caught in the circular reasoning of "it's bigger because it's bigger." Although the responses were not necessarily incorrect, the explanations lacked depth and precision. Further, when the teacher asked him to compare the weights of different objects such as marbles and cubes to balance a scale with a weight on one end, the student was less likely to be able to make statements that reflected the relative weight of the objects (i.e., you require fewer heavier objects than lighter ones) than his general education peers.

Teacher:	Name someone who is taller than you.
Gordon:	My dad.
Teacher:	How do you know this?
Gordon:	Because he's bigger than me.
Teacher:	Name something that weighs more than you.
Gordon:	An elephant.
Teacher:	How do you know this?
Gordon:	Because it's bigger.
Teacher:	Name something that takes up more space than you.
Gordon:	An elephant.
Teacher:	How do you know this?
Gordon:	Because it's bigger than me.

Some students have difficulty using the tools necessary for measuring objects, such as rulers, scales, and balances. Those students with fine-motor problems, such as those with neurological or orthopedic impairments, may be physically unable to manipulate these tools without accommodations or assistance. Also, some students may need extra support to learn the proper techniques for using these tools, such as making sure that the tape lies flat and measuring end to end with the ruler. Some misconceptions about the tools themselves may also lead to errors, as in a misunderstanding of the beginning point. When learning counting, students typically begin at "one" rather than "zero." Although such errors are common for all children, students with learning disabilities who do not think flexibly are especially prone to rely on rote learning of procedures, such as placing the "one" mark at the beginning of the line to be measured (Lehrer 2003) or counting the lines (markings on the ruler) rather than the spaces (distance between the line markings) when measuring (Boulton-Lewis, Wilss, and Mutch 1996; Cannon 1992). The understanding that the distances between intervals on a standard measuring instrument must be equal (e.g., a foot-long ruler is divided into twelve equal intervals, each representing one inch) varies among young children (Petitto 1990) and could be a source of error for students with disabilities.

All these difficulties may interfere with students' abilities to compare and estimate. Students with learning disabilities must gain extensive experience developing and using common referents for measures, such as the meter being about the height of the top of a doorknob or the back of a tall chair.

Finally, many students with disabilities have difficulty applying the knowledge and skills from the classroom to solve problems, sometimes referred to as difficulties with transfer or generalizability of skills to novel contexts. Often, teachers assume that students connect a classroom activity and a real-life application. However, students with disabilities may need teachers to explicitly state, discuss, and illustrate such connections with multiple concrete examples.

General education teachers teaching in inclusive settings often do not get enough information on the learning needs of students with disabilities or on ways to address these needs. Table 8.1 gives a checklist that teachers may find useful in identifying potential concerns and reflecting on ways to work with students who have disabilities (adapted from DeSimone 2003; Tucker, Singleton, and Weaver 2002). The table lists some common characteristics of students with learning disabilities. During instruction, teachers may observe actions on the part of students that could indicate difficulties with learning. Teachers can then use the suggestions to reflect on, or brainstorm with other teachers on, strategies to work with the student to maximize the usefulness of the learning activity.

Table 8.1
Specific learning difficulties related to a given measurement activity

Difficulty with:	Potential behavior indicators	Teaching suggestions
Attending to tasks	Playing with manipulatives as if they were toys rather than engaging in the mathematical task; looking around the room	Have the student take an active role in measuring and recording; quicken pace of lesson for shorter downtime; give only as many materials as necessary; have student self-monitor time on task

Table 8.1—*Continued*

Difficulty with:	Potential behavior indicators	Teaching suggestions
Maintaining attention for the class period	Measuring other children in the room; wandering around	Have the student take notes when others are reporting their work; build short breaks into longer activities
Keeping place on a page in the text or workbook	Losing directions; not knowing where to read; dropping books and pencils in frustration	Give directions for each activity on a separate page; highlight parts most directly relevant to the student with disabilities; assign a "buddy" to help with getting started
Correctly identifying symbols or numerals	Incorrect naming; errors in writing numbers or operational signs; errors copying or writing formulas	Review necessary symbols for the entire class before the activity; check student's work often; ask questions to challenge errors
Using measuring tools	Inaccurate reporting of measurements; incorrect positioning of tools; indicating frustration	Modify tools, such as marking the inch lines on a ruler in blue and 1/2-inch lines in red; review use of tool with entire class before lesson; emphasize key issues such as starting at 0 and counting spaces on ruler
Following a sequence of steps to solution	Errors in solutions and solution processes; guessing at answers; messy paper with many erasures	Supply a reference card with procedures for student to emulate; use mnemonics where appropriate to assist memory; furnish manipulative materials so student can represent the actions and then write the formulas
Oral or written communication in mathematics	Use of "everyday" rather than mathematical vocabulary; avoiding writing or speaking activities in the classroom	Break down writing or speaking requirements into segments; give frequent opportunities for communication; reinforce the use of mathematical vocabulary
Reading	Frustration with reading; inability to read written problems	Teach sight word vocabulary relevant to measurement

Table 8.1—*Continued*

Difficulty with:	Potential behavior indicators	Teaching suggestions
Determining relevant information	Distracted by irrelevant information in written problems	Teach highlighting technique
Interpreting pictures and diagrams	Incorrect labeling of information on a diagram; inability to draw a representation of a figure	Have student verbalize information on diagram; furnish an illustrative model for drawing

Considerations while Making Curricular Adaptations

The following are essential considerations for teachers developing or adapting curricula for students with disabilities:

1. *Integrate topics within learning activities.* Recommendations for instructional or curricular modification for students with disabilities in early elementary-level classes often include decreasing the pace of instruction to give students time to master basic concepts. However, with a decreased instructional pace, students may not be able to learn all the topics they need to complete graduation requirements. Instead, teachers need to develop learning activities that integrate instruction of several topics to avoid compromising curriculum goals. For example, teachers can easily integrate arithmetic computation activities into the instruction of measurement as students determine area, perimeter, volume, and other dimensions. Students can learn the names and characteristics of various shapes as they learn how to measure their attributes. Teachers should also review the curriculum sequences and priorities of the state and school district to ensure addressing important topics early in the school year and not overlooking them.

2. *Incorporate both skill development and reasoning.* Individualized education plan (IEP) recommendations focus on basic skill development (e.g., knowing the names of units used to measure distance, area, volume; learning the formulas for determining the areas of rectangles and circles). Although some repetition and drill in these skills often benefits students with learning, attention, or memory difficulties, such activities should not make up the bulk of instructional time. Teachers should focus on developing mathematical reasoning through presenting problems with real contexts and applications. Many resources exist for problems using measurement, including the NCTM Navigations materials and the NCTM Illuminations Web site.

3. *Plan to phase out assistive materials.* Mathematics teachers are often encouraged to use manipulative materials in instruction. For the topic of measurement, teachers

should distinguish between materials that are tools (e.g., measuring tapes, rulers, balance scales, measuring cups) and those that are aids (e.g., tiles, blocks, fraction strips). Teachers should gradually phase out use of the latter as the students develop facility with abstract reasoning. Otherwise, extensive use of manipulative materials could interfere with, rather than aid, the development of strategic thinking.

4. *Develop mathematical literacy.* Learning activities must include spoken and written mathematical communication, which underscores the need for precision and accuracy. IEPs and instructional planning do not usually address mathematical literacy, even though it is a cornerstone of developing reasoning. Mathematical communication involves using specialized terminology, symbols, and syntax, which students must use with precision (e.g., "area" is not the same as "perimeter"; "greater length" is not "bigger") and appropriate referents (e.g., 3 cm is not equal to 3 inches).

5. *Make explicit links to real-world applications.* Teachers cannot assume that students with learning disabilities transfer classroom learning to other applications. In collaboration with parents, teachers can use information from measurement activities conducted at home (cooking with a recipe, building a birdhouse, traveling from one place to another) to reinforce formal instruction in the classroom. Teachers can refer explicitly to applying measurement in subject areas such as science and social studies, as well as its usefulness in various professions. Such connections enhance the development of mathematical reasoning and increase student motivation to learn.

▼ **Connections to Parents and Caregivers**

Parents and caregivers of children with disabilities often assume responsibility for teaching many mathematics concepts at home, albeit they may not know that they are teaching mathematics. Children often assist their parents in cooking, home renovation, and shopping tasks, all of which involve some form of measurement. Educators can show parents effective ways of involving their children in these family activities while teaching appropriate mathematics concepts. For example, the family is making a double batch of cookies. Rather than the parent doubling the recipe, the child can double the recipe by using problem-solving strategies taught in school. Teachers must communicate to parents the strategies that they are teaching in school, and parents must teach the school staff the family's strategies to approach these tasks.

Instructional Activities and Ideas

Here we present ideas and suggestions for classroom instruction, representing the three major aspects of measurement (units, tools, and comparisons).

Teaching about units

Following is a vignette about measuring with nonstandard units. Mr. Simmons has twenty-three students in his kindergarten classroom. Jonathan and Gordon have learning disabilities, and Samantha is classified as having moderate mental retardation. Mr. Simmons has just demonstrated measuring how many "pencils" long several objects are in the room. The activity for the children is to measure how many "paper clips," "crayons," and "straws" long several objects are (a small box, a book, and a chair) and to report on which units required the most units for measuring the same objects.

Mr. Simmons: Okay, now everyone get into your teams and choose someone to come up and get materials.

Mr. Simmons has assigned the three students with disabilities to work in separate groups so that they may benefit from appropriate models of academic and social behavior and assistance from their classmates without disabilities. The small-group arrangement addresses Jonathan and Gordon's IEP goals of increasing their interpersonal communication skills. Mr. Simmons uses cooperative learning strategies to ensure that each student in the group has understood the task and knows the answer, thereby addressing Jonathan and Gordon's academic needs and keeping them at pace with their classmates.

As the children form their groups and receive materials, consisting of paper clips in a cup and a set of worksheets to complete, Mr. Simmons addresses Samantha's IEP goal of identifying numbers. She receives a worksheet that looks like those the other children have, only with several choices of numerals instead of blank spaces. Her group members are to ensure that Samantha hears what the measurement result is and direct her to circle that answer on her paper. Her worksheet also has the words "most" and "least" written at the bottom, so Samantha can copy the words down after figuring out the answers. Samantha's group, which includes Gretchen and Phil, assists her as follows:

Gretchen: Six, the book is six paper clips long.

Samantha: Six?

Gretchen: Yeah, six, circle six. [*Samantha circles the numeral 6.*]

Phil: Good, that's a 6.

Samantha: Thanks, that's a 6.

Observing, the teacher joins the group and compliments them, gesturing appreciatively at Samantha's correct response. He asks if they can show her that six is correct and guides the group to allow Samantha to count the paper clips. Mr. Simmons makes a note to return and see whether Samantha can perform a similar measurement independently. The activity reinforces the students' counting skills and subtraction (comparing numbers of units). It also develops their mathematical literacy and communication abilities.

As the school year progresses and students become comfortable with using units to measure, Mr. Simmons extends the activity by incorporating problems that involve reasoning and connect measurement to actual applications. The discussion moves toward the importance of standard measurement units.

One student in the class is the designated movie director, who wants to find out the shortest person in the room to play the part of a child in a movie. The teacher designates five teams, each comprising one student who is the actor who gets measured, one student who is the agent, one student who conducts the measurement, and one who records the measurement. The first team receives crayons as the measuring unit; the second team receives straws; the third team gets some paint stirrers; the fourth, shoelaces; and the fifth, a ruler.

Mr. Simmons plans to conduct the activity twice, so that Gordon, Jonathan, and Samantha can each have a turn at being involved in both measurement and communication activities.

After measuring the height of the actor, each team records their units and checks their answers, reinforcing their measurement skills. The actors and agents then confront the movie director with their data (the count written on a sheet of paper), using communication skills to present the information.

Mr. Simmons leads a discussion, asking which team had the fewest units (perhaps shoelaces).

Mr. Simmons: This child is only 4 shoelaces tall, and that one is 8 straws. So the first child is the shortest, right?

The students grapple with the question, and Mr. Simmons furthers the discussion.

Mr. Simmons: But crayons are the smallest measuring unit. Is the child measured with crayons the shortest child? How can the movie director make the correct decision?

As the students use visual inspection as a strategy, Mr. Simmons encourages them to question and verify their answers. He raises questions about using a common referent (standard inch and foot units on the ruler) and precision when two students visually appear to be the same height (fractional units).

The preceding example addressed Gordon and Jonathan's academic and social IEP goals, included Samantha as a fully participating member of the group, emphasized reasoning, and incorporated mathematical communication and literacy.

Teaching the use of tools

Once students have learned the importance and use of measurement, they need to learn about the physical tools of measurement (e.g., rulers, scales) and conceptual tools (e.g., formulas for area and volume) so that they can communicate their understanding of the properties of an object in a way that other individuals can access (McClain et al. 1999). Students often encounter formal or standardized tools for measurement such as rulers, measuring cups, or balance springs/scales around the second grade.

Teachers must especially challenge elementary and middle school students with special needs to learn *concepts* about units, combining units, using rulers, and solving problems. For example, for learning about measuring length, they might measure distance in the classroom by counting paces and then eventually construct a unit of units, such as a "foot strip" consisting of traces of their feet glued to a roll of adding-machine tape (Stephan et al. 2004). Students may then confront the idea of expressing their result in different-sized units (e.g., 15 paces or 3 foot strips, each of which has five paces). They also discuss how to deal with leftover space, to count it as a whole unit or as part of a unit. Measuring with units of units helps students think about length as a composition of these units. Furthermore, it forms the basis for constructing rulers.

Eventually, students should confront more difficult length-related tasks to develop better *measurement sense*. For instance, they could measure perimeters so that they have to iterate units "around a bend." Asking students to make their own drawings and wrestle with a variety of problems such as finding all the rectangles they can whose perimeter is 120 or the type of problem in figure 8.2 ensures that they understand length measurement and its application (Barrett and Clements 2003; Clements and Barrett 1996). They can share their solutions with classmates and discuss the applications of perimeter in building fences or designing garden paths.

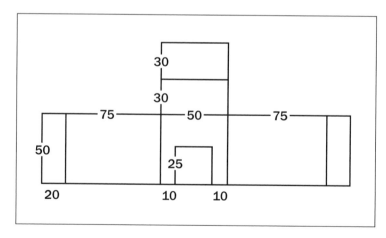

Fig. 8.2. Students fill in all the missing measures (Clements et al. 1995).

For students who do not or cannot segment lines to iterate units and partition lengths, teachers can guide them to continually tie the results of that activity to their counting. For example, they might draw a figure, measure it, and draw it again using the same (and later, a smaller) measure. Students can discuss the importance of scale drawing in a variety of applications such as map drawing and art. Some students with disabilities may benefit by walking around perimeters such as rugs and counting their steps and then connecting that activity to measuring smaller perimeters with manipulatives or rulers.

Formulas are tools used to solve problems. As with many mathematical tools, they are efficient and effective. However, many students apply formulas to obtain answers without meaning (Clements and Battista 1992). They need to build meaning for the mathematical concepts first. For example, although most elementary-level students have no difficulty covering a region with tiles and finding area by counting individual tiles, many cannot *represent* the results of such actions in a drawing. These children do not interpret arrays as rows and columns, which obstructs their learning about area measurement and renders formulas meaningless (Outhred and Mitchelmore 1992). Learning formulas for area and volume measurement thus begins with learning about arrays.

▼
Phasing Out Assistive Materials

As children begin more formal measurement activities, teachers need to develop a plan for managing a transition from concrete materials to paper–pencil representation, using appropriate symbols and mathematical conventions. To help students estimate usefully, teachers may give them some easy personal referents for commonly used measures. For example, an inch may be approximately the length of the thumb for a five- to six-year-old child, a yard may be approximately the length from the tips of the fingers in outstretched hands, and a meter may be the distance from the floor to the top of a doorknob. Having such referents will reduce the confusion and stress that a student with disabilities might experience when first facing a measurement task. Teachers can then link the symbols to referents that children are comfortable and familiar with, giving a better chance for developing a correct understanding. Teachers can present estimation activities using such personal referents to students, with encouragement to use the appropriate mathematical vocabulary, until personal referents are no longer necessary.

Although we identified no research that specifically examined students with disabilities, research with general education students reveals that several misconceptions hamper students attempting area problems. These include (1) confusing area and perimeter, (2) applying the formula for the area of a rectangle to other plane figures, (3) believing that doubling sides of a square doubles its area, (4) poor understanding of the relationship between addition and multiplication, (5) misunderstanding of the structure of a mathematical array, and (6) applying formulas without connecting them to the presence of square units within a figure (Outhred and Mitchelmore 2000).

Structuring an array is particularly important for students with special needs. As previously noted, most students cannot "see" an area as composed of arrays with rows and columns, or three-dimensional space composed of, say, cubes in layers. Researchers have identified levels of development in the understanding of rectangular space. These levels are useful guides for teachers in setting realistic instructional goals and moving students toward more advanced learning (Battista et al. 1998).

- Level 1: little or no ability to organize, coordinate, and structure two-dimensional space (cannot represent covering a rectangle with tiles without overlaps or gaps)

- Level 2: complete covering, but counting incorrectly (cannot keep track of which units were counted, e.g., counts around the border and then unsystematically counts internal units)

- Level 3: covering and counting but again with no row or column structuring

- Level 4: the local, incomplete use of rows or columns (e.g., counts some, but not all, rows as a unit)

- Level 5: structuring the rectangle as a set of rows

- Level 6: iterating those rows (e.g., counting each row of 5, "5, 10, 15 . . . " in coordination with the number of squares in a column (e.g., counting by 5)

- Level 7: understanding that the rectangle's dimensions give the number of squares in rows and columns and thus meaningfully calculating area from these dimensions

Structuring arrays is especially difficult for children with perceptual or visual difficulties. They will need additional time and experience and often physical guidance in making arrays, such as using inch grid paper and inch tiles (e.g., the square from pattern blocks), or even adaptive materials such as plastic grids. One-inch graph paper is readily available online.

When giving students activities to determine area, teachers commonly give tiles or other manipulatives that children can assemble within an outline and then count. This instructional activity has some merit as a beginning activity but can also promote misconceptions, particularly with students who have difficulty with transfer from concrete to more abstract representations and those who have not internalized the importance of precision in mathematical communication. The ease of using the tiles can actually become a barrier to learning (Hiebert and Carpenter 1992), because the activity supports using a basic counting strategy with many single units rather than developing an understanding that one must iterate a unit, or combine it into "units of units" such as foot strips or rows of units, as described previously. Such more sophisticated understandings are the basis for more advanced reasoning about measurement and applying formulas.

To build a strong understanding of the concept of area, we suggest asking students to reason from several perspectives, thus challenging their misconceptions. For example, to determine area students might start by comparing regions of different shapes and sizes drawn on graph paper, directly and by partitioning them. Teachers may then guide them to use one partition as a unit to measure the others (Lehrer 2003). Students may then be offered multiple units to determine area. They should realize that there are to be no gaps or overlapping units and that the entire region should be covered. See figure 8.3 (taken from *A Research Companion to "Principles and Standards for School Mathematics"* [NCTM 2003, p. 186]).

Next, they should learn to structure arrays. Figuring out how many squares are in pictures of arrays, with less and less graphic information of clues, is an excellent task (Battista et al. 1998). Students can initially tile rectangular regions and keep count. Then they might be given a single tile to place or draw repeatedly and asked to draw the results of their covering (cf. Reynolds and Wheatley 1996). This approach would allow the teacher to assess whether the students understand the concept of an array and whether they can accurately construct an array where units do not overlap, are disassociated, or are of unequal sizes. For example, some students drew a series of square tiles within the region they were measuring, yet

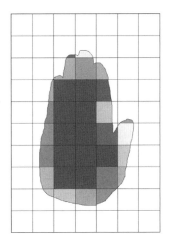

Fig. 8.3. Measuring area

there were obvious gaps between tiles. Other students drew an array that had an unequal number of units in each row. Students need tasks and instruction that leads them through the levels of learning this structuring (Battista et al. 1998).

Next, teachers might ask students given an array to draw the outline of a figure on a separate sheet of paper. Doing so would determine whether students could apply their knowledge of multiplication through arrays as learned in arithmetic to a geometric representation. At this point, discussion should highlight that the length of the sides of a rectangle can determine the number of units in each row and the number of rows in the array. Next, children can meaningfully learn to multiply the two dimensions as a shortcut for determining the total number of squares. An appropriate understanding of linear measurement (that the length of a side specifies the number of unit lengths that will fit along it) is essential. Only then can students construct the area formula.

For an assessment, a teacher could draw a unit (e.g., inch square) on paper with a larger figure, and students must determine the area of the larger figure without manipulatives. An understanding of children's ability to use the formula effectively would help the teacher address issues such as attention to only one dimension within an array; misapplication of the formula; and whether computation errors, not conceptual understanding, are at the root of any incorrect answers.

For students with severe memory difficulties related to attention deficits, traumatic brain injury, or learning disabilities, some supports may be necessary in addition to activities that build conceptual understanding. Mnemonic strategies can help students to associate formulas with concepts. One mnemonic for the formula for area of a circle (πr^2) is picturing a square pie floating in the air. The "air" reminds students that the formula is for area. The square pie reminds students of the formula. Similarly, a mnemonic for remembering the formula for the circumference ($2\pi r$) is the picture of two circular pies. "Circular" reminds the student that the formula is for circumference because the beginning of both words is the same.

Another important element is for special education teachers to collaborate with general education teachers so that what the students learn in their resource rooms or with their aides effectively supports their classroom instruction. For example, if Jonathan from the earlier illustration needs extra help learning the terms *diameter*, *circumference*, and *area* before the lesson, the special education teacher can give extra instructional time with the relevant vocabulary, so that Jonathan could focus on the essential mathematics during the regular mathematics lesson.

When Nita, who has developmental disabilities, is also included in the upper elementary grades, the special education teacher could give her modified worksheets with a diagram of the figure to be measured constructed on graph paper. This way, she can use a counting strategy and break down the steps of the formula so she can practice addition or multiplication, as per her IEP goal. Yet she has the correct answers and can participate if called on in class (see earlier list of levels of development in the understanding of rectangular space).

Teachers can use strategies similar to those described earlier for area to teach volume as students move from upper elementary grades to middle school. That students learn to structure three-dimensional space as arrays is even more essential, because they cannot see all the units that they need to count.

At early stages in learning about volume, for example, students count only the faces of cubes in a cubic building, or they attempt to count cubes but forget some internal ones and double count cubes at vertices or edges (Battista and Clements 1996). Eventually, students build a layered structure, in which they determine the number of cubes in each layer and then compute the total number of cubes by skip counting or multiplying this number by the number of layers (the height).

Multiple activities, including building, drawing, completing, and representing arrays, are usually needed. For example, teachers might ask students to (1) make boxes out of nets (two-dimensional patterns that fold into three-dimensional shapes); (2) exchange their nets; (3) before they rebuild the box from the net they were given, determine how many cubes will fit in the box; and (4) reconstructing the box and checking their prediction. This task involves not only direct manipulation of concrete materials but also problem solving and prediction based on mathematical reasoning (especially in step 3), with the concrete materials used (in step 4) to test their prediction. This approach illustrates using manipulatives to best advantage, minimizing any disadvantages. (For more information and ideas for tasks, see Battista and Clements 1996, 1998; Battista et al. 1998.)

Teaching about conversions of units

In the middle grades, measurement activities increasingly incorporate using formulas to determine first-order measures (e.g., area) or second-order measures (e.g., rate). In addition to difficulties with concepts of measurement, arithmetic computation difficulties may also contribute to errors in performance. Because this book has previously discussed number and operations (chapter 5) and algebra (chapter 6), we will simply note that one must consider those same issues when teaching measurement. Ratio and proportion are also important aspects of measurement, which this chapter discussed earlier, in the context of fractions, and therefore we need not discuss them extensively here.

Figure 8.4 reflects potential confusion with the many scales in the American customary measurement system. Although several sources have argued in favor of the metric system for its consistency in using base-ten numbers across various types of measures, widespread acceptance in the United States is unlikely. Thus, schools will continue teaching both systems.

In figure 8.4, those items not requiring renaming are simple single-digit addition tasks. However, in each example with renaming, the base is a different value. A student must not

(1)

4 lb, 8 oz

+ 3 lb, 6 oz

(2)

4 lb, 8 oz

+ 3 lb, 9 oz

(3)

2 weeks, 5 days

+ 2 weeks, 1 day

(4)

2 weeks, 5 days

+ 2 weeks, 8 days

(5)

3' 7"

+ 5' 3"

(6)

3' 7"

+ 5' 8"

Fig. 8.4. Example of conversion of units

only correctly rename but also remember to do so at different times for each task. Potential errors may include (1) not renaming at all (e.g., deriving 7 lb, 17 oz for item 2), (2) using base ten for all renaming, (3) using incorrect bases (e.g., 7 for pounds, 12 for days in a week), (4) ignoring the various units and adding all the numbers up, and (5) getting confused by the symbols representing units and refusing to do the task. For students who have experienced past difficulties with reading and understanding symbols for addition, subtraction, and so forth, the difficulties are compounded. Teachers, therefore, need to find ways to educate students about the various bases that customary measures use and to capitalize on their prior knowledge so they can effectively apply it to calculations.

The customary measurement system's different bases are somewhat arbitrary; however, their development is based on historical events and usage. To promote student reasoning, teachers could discuss the historical origins of the systems. For example, in ancient times round coins were minted that could be divided into halves, then fourths, then eighths, and then sixteenths, giving the basis for the sixteen-ounce "pound."

Another reasoning activity could involve looking at popular publications and media, such as newspapers, television, trade books, or even textbooks, to find incorrect representations. For example, some publishers use 10.5 years to refer (correctly) to ten years and six months, but others have used 10.5 to refer (incorrectly) to ten years and five months. Other points of discussion may be determining what constitutes half a week or how to determine whether the symbols ' and " refer, respectively, to feet and inches or pounds and ounces in a given context.

To develop fluency with conversion formulas, teachers must offer many opportunities for practice, since memory difficulties are often a source of errors for students with disabilities. Manipulative materials such as Cuisenaire rods or Unifix cubes may be used along with a board such as that for teaching place value to younger children. Teachers can present regrouping activities first with manipulatives, then with symbols or marks representing manipulatives, and finally with formal symbolic representations.

Developing a conceptual understanding of conversion first requires that students understand scales of measurement. For example, learning activities related to linear measurement should incorporate discussions such as the following: What is an appropriate measurement unit to measure the distance traveled between two cities (miles)? Between two houses on a street (feet)? Between two windows on a wall (inches)?

Such discussion gives students an understanding of the logic behind different scales of measurement. It also helps them develop a foundation for making reasonable estimates of distance and places to use rounding of quantities (e.g., distances between cities are often rounded to the nearest tens unit in conversation or news reporting). Students with disabilities often have difficulty visualizing information because they lack understanding of what scales to use; they also require practice in developing the necessary vocabulary to express their understanding of measures.

Problem-solving activities can initially use one unit (feet) and then move to a combination of units (feet and inches). When students actually measure using maps or blueprints, teachers can discuss the concept of "scale," since one inch on a map may represent several miles. This concept is another application of conversion, where a given unit represents another unit, and students must multiply by the scale (using an understanding of ratio) to determine the correct answer.

▼   **Developmental Progression in Measuring and Estimating**

Research with young children without disabilities points to a developmental progression in the ability to accurately measure and estimate, on the basis of children's facility with the Piagetian concept of conservation (see Miller 1989 for a discussion and examples). A similar developmental pattern may be hypothesized for students with learning disabilities, with some variations (Riley 1989). At the age of three years, children rarely attempt to verify their comparative statements about length (taller/shorter), though by the ages of four and five, they often use visual comparison to create equivalencies and even use rudimentary tools, such as their own arm lengths, to make quantitative judgments (Kamii and Clark 1997). By the age of about seven, they begin to use standardized tools for measurement.

From the preschool and kindergarten years through primary school, teachers should check that every child can compare lengths and order several lengths. Along with classifying and conservation, this essential skill may underlie broad domains of mathematical and general thinking (Pasnak 1987; Pasnak et al. 1996).

Reflections on Teaching Measurement: Questions for Discussion

1. In what ways are measurement tasks challenging specifically for students with disabilities?

2. What general supports can benefit students with disabilities in learning measurement concepts and skills? What are some specific supports related to measuring length, area, volume, and weight?

3. Why should teachers of students with disabilities use strategies other than slowing the pace of instruction and supplying manipulatives?

4. How can analyzing student errors in measuring or student behavior assist a teacher in designing instruction appropriate for students with disabilities?

References

Barrett, Jeffrey E., and Douglas H. Clements. "Quantifying Path Length: Fourth-Grade Children's Developing Abstractions for Linear Measurement." *Cognition and Instruction* 21 (December 2003): 475–520.

Battista, Michael T., and Douglas H. Clements. "Students' Understanding of Three-Dimensional Rectangular Arrays of Cubes." *Journal for Research in Mathematics Education* 27 (May 1996): 258–92.

———. "Students' Understanding of Three-Dimensional Cube Arrays: Findings from a Research and Curriculum Development Project." In *Designing Learning Environments for Developing Understanding of Geometry and Space*, edited by Richard Lehrer and Daniel Chazan, pp. 227–48. Mahwah, N.J.: Lawrence Erlbaum, 1998.

Battista, Michael T., Douglas H. Clements, Judy Arnoff, Kathryn Battista, and Caroline Van Auken Borrow. "Students' Spatial Structuring of 2D Arrays of Squares." *Journal for Research in Mathematics Education* 29 (November 1998): 503–32.

Boulton-Lewis, Gillian M., Lynn A. Wilss, and Sue L. Mutch. "An Analysis of Young Children's Strategies and Use of Devices of Length Measurement." *Journal of Mathematical Behavior* 15 (September 1996): 329–47.

Cannon, P. L. "Middle Grade Students' Representations of Linear Units." In *Proceedings of the Sixteenth Psychology of Mathematics Education Conference*, edited by William Geeslin and K. Graham, vol. I, pp. 105–12. Durham, N.H.: Program Committee of the Sixteenth PME Conference, 1992.

Clements, Douglas H., and Jeffrey Barrett. "Representing, Connecting and Restructuring Knowledge: A Micro-Genetic Analysis of a Child's Learning in an Open-Ended Task Involving Perimeter, Paths and Polygons." In *Proceedings of the Eighteenth Annual Meeting of the North America Chapter of the International Group for the Psychology of Mathematics Education*, edited by E. Jakubowski, D. Watkins, and H. Biske, vol. I, pp. 211–16. Columbus, Ohio: ERIC Clearinghouse for Science, Mathematics, and Environmental Education, 1996.

Clements, Douglas H., and Michael T. Battista. "Geometry and Spatial Reasoning." In *Handbook of Research on Mathematics Teaching and Learning*, edited by Douglas A. Grouws, pp. 420–64. New York: Macmillan, 1992.

Clements, Douglas H., Michael T. Battista, Joan Akers, Virginia Woolley, Julie S. Meredith, and Sue McMillen. *Turtle Paths—2-D Geometry*. Cambridge, Mass.: Dale Seymour Publications, 1995.

Council of Chief State School Officers (CCSSO). *Common Core State Standards for Mathematics*. Washington, D.C.: CCSSO and the National Governors Association Center for Best Practices, 2010. www.corestandards.org.

DeSimone, Janet. "Middle School Mathematics Teachers' Perceptions of Included Students." Ed.D. diss., St. John's University, 2003.

Forrester, M. A., Janette Latham, and Beatrice Shire. "Exploring Estimation in Young Primary School Children." *Educational Psychology* 10, no. 4 (1990): 283–300.

Fuchs, Lynn S., and Douglas Fuchs. "Principles for the Prevention and Intervention of Mathematics Difficulties." *Learning Disabilities Research and Practice* 16 (May 2001): 85–95.

Hiebert, James, and Thomas P. Carpenter. "Learning and Teaching with Understanding." In *Handbook of Research on Mathematics Teaching and Learning*, edited by Douglas A. Grouws. New York: MacMillan, 1992.

Kamii, Constance, and Faye B. Clark. "Measurement of Length: The Need for a Better Approach to Teaching." *School Science and Mathematics* 97 (March 1997): 116–21.

Lehrer, Richard. "Developing Understanding of Measurement." In *A Research Companion to "Principles and Standards for School Mathematics,"* edited by Jeremy Kilpatrick, W. Gary Martin, and Deborah Schifter, pp. 179–92. Reston, Va.: National Council of Teachers of Mathematics, 2003.

McClain, Kay, Paul Cobb, Koeno Gravemeijer, and Beth Estes. "Developing Mathematical Reasoning within the Context of Measurement." In *Developing Mathematical Reasoning in Grades K–12*, 1999 Yearbook of the National Council of Teachers of Mathematics (NCTM), edited by Lee V. Stiff and Frances R. Curcio, pp. 93–106. Reston, Va.: NCTM, 1999.

Miller, Kevin F. "Measurement as a Tool for Thought: The Role of Measuring Procedures in Children's Understanding of Quantitative Invariance." *Developmental Psychology* 25 (July 1989): 589–600.

National Council of Teachers of Mathematics (NCTM). *Principles and Standards for School Mathematics*. Reston, Va.: NCTM, 2000.

———. *A Research Companion to "Principles and Standards for School Mathematics."* Reston, Va.: NCTM, 2003.

———. *Curriculum Focal Points for Prekindergarten through Grade 8 Mathematics: A Quest for Coherence.* Reston, Va.: NCTM, 2006.

Outhred, Lynne N., and Michael C. Mitchelmore. "Representation of Area: A Pictorial Perspective." In *Proceedings of the Sixteenth Psychology in Mathematics Education Conference*, edited by William Geeslin and K. Graham, vol. II, pp. 194–201. Durham, N.H.: Program Committee of the Sixteenth PME Conference, 1992.

————. "Young Children's Intuitive Understanding of Rectangular Area Measurement." *Journal for Research in Mathematics Education* 31 (March 2000): 144–67.

Pasnak, Robert. "Acceleration of Cognitive Development of Kindergartners." *Psychology in the Schools* 24 (October 1987): 358–63.

Pasnak, Robert, S. E. Madden, Valerie A. Malabonga, and R. Holt. "Persistence of Gains from Instruction in Classification, Seriation, and Conservation." *Journal of Educational Research* 90 (November–December 1996): 87–92.

Petitto, Andrea L. "Development of Numberline and Measurement Concepts." *Cognition and Instruction* 7 (March 1990): 55–78.

Reynolds, Anne, and Grayson H. Wheatley. "Elementary Students' Construction and Coordination of Units in an Area Setting." *Journal for Research in Mathematics Education* 27 (November 1996): 564–81.

Riley, Nancy J. "Piagetian Cognitive Functioning in Students with Learning Disabilities." *Journal of Learning Disabilities* 22 (August–September 1989): 444–51.

Stephan, Michelle, Janet Bowers, Paul Cobb, and Koeno P. E. Gravemeijer. "Supporting Students' Development of Measuring Conceptions: Analyzing Students' Learning in Social Context." *Journal for Research in Mathematics Education Monograph 12*. Reston, Va.: National Council of Teachers of Mathematics, 2004.

Tucker, Benny F., Ann H. Singleton, and Terry L. Weaver. *Teaching Mathematics to All Children: Designing and Adapting Instruction to Meet the Needs of Diverse Learners*. Upper Saddle River, N.J.: Pearson, 2002.

CHAPTER 9

Data Analysis and Probability

Cynthia W. Langrall
and Edward S. Mooney

In recent years, data analysis and probability have received increased attention in the mathematics curriculum. The focus on these topics has moved beyond the middle school and secondary levels and has addressed the ongoing development of students' understanding of statistics and probability, beginning as early as the primary grades. This increased attention reflects the importance of statistics and probability in our daily lives, that is, the need for all citizens to develop the conceptual and practical tools for living and working in an information-based society (Shaughnessy, Garfield, and Greer 1996). Data analysis and probability have become important aspects of mathematical learning for all students, including those with special needs.

Expectation and Needs: Data Analysis and Probability Standard

According to the *Principles and Standards for School Mathematics* (National Council of Teachers of Mathematics [NCTM] 2000, p. 48), pre-K–12 instructional programs should enable all students to formulate questions that can be addressed with data and collect, organize, and display relevant data to answer them; select and use appropriate statistical methods to analyze data; develop and evaluate inferences and predictions that are based on data; and understand and apply basic concepts of probability. *Curriculum Focal Points for Prekindergarten through Grade 8 Mathematics: A Quest for Coherence* (NCTM 2006) reiterated these expectations. More recently, the *Common Core State Standards Initiative* (Council of Chief State School Officers [CCSSO] 2010) identified statistics and probability as crucial areas for instruction beginning in grade 6.

The seemingly endless opportunities for students to collect and analyze information call for rich curricular experiences in data analysis and probability that go beyond simply reading information from preconstructed graphs or determining probabilities on the basis of a listing of outcomes. Students must be actively engaged in collecting and analyzing data and investigating probabilistic situations (Franklin et al. 2007; NCTM 2000).

Rationale

Proficiency in statistical and probabilistic reasoning enables students to become productive, participating citizens in an information-based society. People encounter data daily in the form of surveys, polls, graphs, and news facts. Many everyday activities require interpreting and using such data. For example, comparison shopping requires the consumer to collect and analyze information from a variety of products and to make decisions from this analysis. People often must make decisions on the basis of assumptions that involve a degree of uncertainty.

Academically, understanding data analysis and probability allows students to make meaningful connections to other branches of mathematics (i.e., number, algebra, measurement, and geometry) as well as connections to other disciplines. However, reasoning in data analysis and probability situations is not necessarily intuitive. Misconceptions about probability and statistics are common among many adults, as well as students. These facts underscore the importance of giving all students instructional experiences aimed at developing understanding of concepts and skills associated with data analysis and probability.

Instructional perspective

Most special needs students will benefit from instruction in data analysis and probability that follows the basic principles of teaching and learning that apply to all children. National Research Council findings support this belief. A report of that study, *Adding It Up: Helping Children Learn Mathematics* (Kilpatrick, Swafford, and Findell 2001), explained that it is in the best interest of special needs students "to assume that the following principles apply to *all* children: (a) learning with understanding involves connecting and organizing knowledge; (b) learning builds on what children already know; and (c) formal school instruction should take advantage of children's informal everyday knowledge of mathematics" (p. 342). They also noted that the learning difficulties of special needs students may stem from the following common mistakes in instruction: "(a) not assessing, fostering, or building on [students'] informal knowledge; (b) overly abstract instruction that proceeds too quickly; and (c) instruction that relies on memorizing mathematics by rote" (p. 342). Although these basic principles of learning apply to all students, it does not follow that all students learn the same way. Rather, these principles recognize that learners, especially those with special needs, are unique in the experiences and knowledge they bring to the instructional setting.

For teachers to understand and build on students' informal knowledge, they must be able to assess the thinking of students in their classrooms, as well as have some knowledge of the general development of understanding within the particular mathematical domain. In the domains of data analysis and probability, such knowledge has only just begun to emerge and then not always in a coherent form that teachers can easily access.

Consequently, most teachers are not familiar with cognitive models that characterize the development of students' statistical and probabilistic thinking.

To address this need, we have developed cognitive frameworks that characterize the development of elementary and middle school students' thinking in both statistics and probability (see fig. 9.1 and 9.8). Our work in constructing these frameworks has taken place in inclusion classrooms and has involved children of all levels of ability and need. We have found these frameworks useful tools for planning, implementing, and evaluating classroom instruction and believe that teachers in a variety of instructional settings can use them effectively.

In the sections that follow, we describe both frameworks and suggest how teachers can use them to guide instruction. First, we explain the theoretical perspective on which the frameworks were constructed. Next, we present each framework, followed by vignettes that illustrate the different levels of thinking that elementary and middle school students, including those with special needs, typically exhibit. We then present issues pertaining to the teaching and learning of data analysis and probability and other connections to practice. The chapter closes with suggestions for using the frameworks to guide the instruction of all students.

Frameworks Characterizing Students' Statistical and Probabilistic Thinking

The frameworks we have constructed to characterize students' statistical and probabilistic thinking are grounded in a twofold theoretical perspective. First, our extensive observations and interviews with students in grades 1–8 found that one can characterize students' thinking as developing across four levels that reflect shifts in the complexity of their thinking. Second, on the basis of a synthesis of the research on statistical and probabilistic thinking, we have concluded that data analysis and probability are multifaceted content areas. They comprise key processes or concepts for which students develop an understanding over time through experiences both within and beyond the classroom.

Levels of thinking

One can characterize students' statistical and probabilistic thinking across four levels: (1) idiosyncratic/subjective, (2) transitional, (3) quantitative, and (4) analytical/numerical. At the idiosyncratic/subjective level, students' thinking is narrowly and consistently bound to idiosyncratic or subjective reasoning unrelated to the given data and often focused on personal experiences or subjective beliefs. Irrelevant aspects of a problem situation may distract or mislead students' reasoning at this level. At the second level, students' thinking is transitional: they begin to recognize the importance of thinking quantitatively, but they apply such reasoning inconsistently. Students reasoning at this level engage in tasks in a relevant way but generally focus only on one aspect of the problem situation. At the third level, students' thinking is consistently quantitative: they can identify the mathematical ideas of the problem situation, and irrelevant aspects do not distract or mislead them.

However, students reasoning at this level do not necessarily integrate these relevant mathematical ideas when engaging in the task. At the analytical/numerical level, students' thinking is based on making connections between the multiple aspects of a problem situation. Students reasoning at this level can integrate the relevant aspects of a task into a meaningful structure (e.g., assigning a probability to an event, creating an appropriate data display, or making a reasonable prediction).

The shifts in complexity of thinking that our framework describes are consistent with the Biggs and Collis (1991) learning cycles and levels of cognitive functioning. Their general model of development describes five modes of functioning: sensorimotor, iconic, concrete symbolic, formal, and postformal. Within each, Biggs and Collis assume three cognitive levels (unistructural, multistructural, and relational) that cycle through each mode and a fourth level (prestructural) that represents thinking not characteristic of the mode. The four levels of thinking in our frameworks correlate with those of Biggs and Collis (table 9.1).

Table 9.1
Levels of thinking

Level of thinking		
General developmental model (Biggs and Collis 1991)	**Probability/ statistics framework**	**Description**
Prestructural	Subjective/ idiosyncratic	Subjective reasoning: students do not focus on relevant aspects of the task
Unistructural	Transitional	Does address a relevant aspect of the task but is often inconsistent; students may revert to subjective reasoning
Multistructural	Quantitative	Based on more than one relevant aspect of the task but may not draw connections between these aspects
Relational	Numerical/ analytical	Based on connections between relevant aspects of the task

In developing our frameworks and using them in instructional settings, we have found that students in the elementary and middle grades typically function in the concrete symbolic mode. However, students sometimes respond at the idiosyncratic/subjective level, which represents thinking characteristic of the iconic mode.

Our frameworks' thinking levels are not labels, nor should one interpret thinking-level descriptors as objectives for students to master to move to higher levels of thinking. Rather, they give teachers knowledge of the general patterns of growth that students typically exhibit, enabling teachers to monitor the development of students' statistical and probabilistic thinking. In our work in inclusion classrooms, we have found that all students, including those with special needs, exhibit thinking across the levels of our frameworks.

Statistical thinking framework

The statistical thinking framework (fig. 9.1) characterizes students' thinking across the four levels of development for the four statistical processes of describing, organizing, representing, and analyzing data. The statistical processes coincide with key elements of handling data that Shaughnessy, Garfield, and Greer (1996) identified. According to these authors, data handling involves "organizing, describing, representing, and analyzing data, with a heavy reliance on visual displays such as diagrams, graphs, charts, and plots" (p. 205). One can therefore view statistical thinking as the thinking that occurs while performing these statistical processes.

Describing data entails the explicit reading of data appearing in tables, charts, or graphical representations. Curcio (1987) considers "reading the data" the initial stage of interpreting and analyzing data. The ability to read data displays becomes the basis on which students make predictions and discover trends. Two *subprocesses* relate to describing data: (1) showing awareness of display features and (2) identifying units of data values.

Organizing data involves arranging, categorizing, or consolidating data into a summary. As with the ability to describe data displays, the ability to organize data is essential for learning how to analyze and interpret data. Arranging data in clusters or groups can illuminate patterns or trends in the data. Measures of center (i.e., mean, median, and mode) and dispersion are useful in comparing sets of data. Organizing data comprises three subprocesses: (1) grouping data, (2) describing typical or representative values of the data, and (3) describing the spread of data.

Representing data involves displaying data graphically. Friel, Curcio, and Bright (2001) stated that the graphical sense involved in representing data "includes a consideration of what is involved in constructing graphs as tools for structuring data and, more important, what is the optimal choice for a graph in a given situation" (p. 145). Representing data, like the previous two processes, is important in analyzing and interpreting data. The type of display used and how the data are represented will determine the trends and predictions that students can make. Also, different displays can communicate different ideas about the same data. Two subprocesses that underlie representing data are (1) constructing a data display for a given data set and (2) evaluating the effectiveness of data displays in representing data.

Analyzing data consists of identifying trends and making inferences or predictions from a data display. Curcio (1987) described two approaches to analyzing and interpreting data: *reading between the data* and *reading beyond the data*. Reading between the data involves making comparisons within the data. Reading beyond the data entails making extensions, predictions, or inferences from the data. The three subprocesses for analyzing and interpreting data include (1) comparing data, (2) making predictions or inferences, and (3) using relative thinking.

The four statistical processes, though we examine them separately, are closely linked. Data analysis builds on reading, organizing, and displaying data.

Statistical thinking framework

		Level 1, idiosyncratic	Level 2, transitional	Level 3, quantitative	Level 4, analytical
Describing data	In examining a data display, demonstrates . . .	Little awareness of the display features	Some awareness of the display features	Awareness of display features	Awareness of relevant display features
	In describing data, identifies . . .	None of the units of the data values	Units of data values incorrectly	Units of specific data values	Units of data values in general
Organizing data	Groups data in a . . .	[Student does not group data]	Nonsummative form	Summative form or by creating new categories or clusters	Summative form by creating new categories or clusters
	Describes the typicalness of data by using . . .	[Student does not describe typicalness]	Partially valid measures	Valid invented measures	Valid measures of center
	Describes the spread of data by using . . .	[Student does not describe spread]	Partially valid measures	Valid invented measures	Valid measures of spread
Representing data	Constructs a data display that is . . .	Incomplete or unrepresentative of the data	Partially complete or partially representative of the data	Complete and representative	Complete, representative, and appropriate
	Evaluates the effectiveness of a data display to represent data based on . . .	Irrelevant features or reasons	Relevant display features	Relevant display features and some reference to the context of the data	Relevant display features and the context of the data
Analyzing data	Compares data with . . .	Incorrect or no comparisons of the data	One comparison or partially correct set of data comparisons	Local or global comparisons of the data	Local and global comparisons of the data
	Makes predictions or inferences that are . . .	Not based on the data	Partially based on the data	Primarily based on the data	Reasonably based on the data and context
	Uses relative thinking . . .	In no analysis of data	Qualitatively	Quantitatively, but not in a reasonable manner	Quantitatively in a reasonable manner

Fig. 9.1. Statistical thinking framework

Examples of students' thinking in data analysis

The following two vignettes illustrate the four levels of students' thinking in data analysis that we saw in our work with elementary and middle school students (Jones et al. 2000, 2001; Mooney 2002). The first vignette focuses on the strategies that students use to organize and represent data. The second vignette illustrates the levels of thinking that students exhibit when they describe and analyze data.

Vignette 1: Organizing and representing data

In this vignette, we present the work of four groups of students to illustrate four levels of thinking with respect to organizing and representing data. These excerpts are from an upper elementary class engaged in the teachers' pets task (fig. 9.2). The teacher introduced the task and asked the students to be prepared to present their data displays to the class. Students worked in pairs, using markers and newsprint to construct the displays. To make the problem accessible to students at a variety of levels of thinking, the teacher made the data available in two forms: as a printed list and on individual cards for students to manipulate. As the students worked, the teacher circulated among the groups, noting the different levels of students' thinking.

Group 1: Paul and Lisa

Paul: We're listing which pets they have. And we added a title. [*See fig. 9.3.*]

Teacher: What information does your display tell you?

Paul: It shows the kind of pets teachers have.

Teacher: Did you think about using the cards to help you organize the data?

Lisa: No, the cards didn't really help. We liked the list the way it was.

Task	The teachers at our school were asked what kinds of pets, and how many of each kind, they have at home. In all, the teachers had 39 pets.
Organizing and representing data	1. Here is a list of the responses teachers gave to the survey (see below). Your job is to present the data to show which pet is the most popular among the teachers.

Doberman	Goldfish	Siamese cat	Poodle
Parakeet	Siamese cat	Burmese cat	Parakeet
Persian cat	Garter snake	Chameleon	Goldfish
Burmese cat	Parrot	Parakeet	Parrot
Collie	Parakeet	Persian cat	Iguana
Poodle	Iguana	Collie	Iguana
Angelfish	Persian cat	Goldfish	Persian cat
Parakeet	Angelfish	Goldfish	Goldfish

Fig. 9.2. Statistical thinking tasks: teachers' pets—
Continues

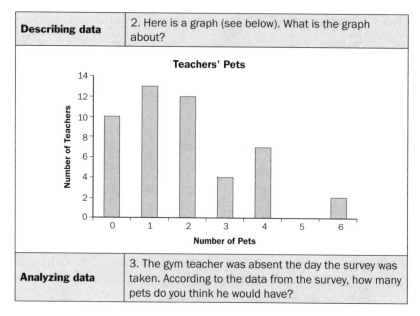

Describing data	2. Here is a graph (see below). What is the graph about?

Teachers' Pets

Analyzing data	3. The gym teacher was absent the day the survey was taken. According to the data from the survey, how many pets do you think he would have?

Fig. 9.2. Statistical thinking tasks: teachers' pets—*Continued*

Fig. 9.3. Paul and Lisa's list of teachers' pets

Group 2: Terrill and Jonathan

Jonathan:　　Terrill stacked up the cards and read off each name. I wrote the name and put an *X* above the name. [*See fig. 9.4.*]

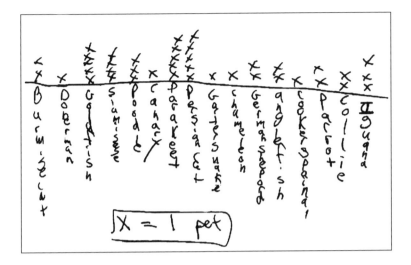

Fig. 9.4. Terrill and Jonathan's line plot

Group 3: *Rosa and Daya*

Daya: We sorted the cards into piles by type of animal, but we didn't know what to do with these [*garter snake, iguana, and chameleon*].

Rosa: Then we wrote the type of animal and drew that many *X*s to make a graph. [*See fig. 9.5.*]

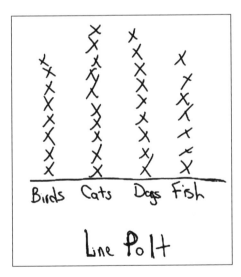

Fig. 9.5. Rosa and Daya's line plot.

Group 4: *Moremi and Chris*

Moremi: We noticed that we could group the animals by cats, dogs, birds, fish, and reptiles. So we made a chart that shows each of the different kinds of pets and the totals for each. [*See fig. 9.6.*]

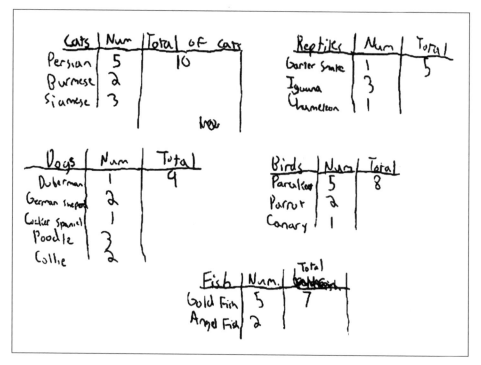

Fig. 9.6. Moremi and Chris's table display

The display that Paul and Lisa (group 1, fig. 9.3) constructed, and their responses to the teacher's questions, represent idiosyncratic thinking for organizing and representing data. Not only did they fail to group the data in any way, but they also failed to construct a display. They simply copied the list of given data.

Terrill and Jonathan's display (group 2, fig. 9.4) illustrates a quantitative level for representing data because their display is both complete and representative of the data. However, their method of organization reflects only a transitional level because the data are grouped but not in a summative form. By way of contrast, the line plot that Rosa and Daya created (group 3, fig. 9.5) characterizes a quantitative level of organizing data because of the categories that they created to group the pets. However, the representation is transitional because they could not determine a category for "other" pets (i.e., garter snake, iguana, and chameleon), making their display unrepresentative of all the data collected.

Moremi and Chris's display (group 4, fig. 9.6) characterizes analytical thinking for both organizing and representing data. They accurately represent the data in a summative form, using new categories. Their table illustrates the idea that data displays or representations are not limited to graphs.

Vignette 2: Describing and analyzing data

In this scenario, a middle school class is working on parts two and three of the teachers' pet task, which relate to the number of pets that each teacher owns. The teacher presented the data display (fig. 9.2) and asked the students to write about the graph in their journals.

Following are excerpts from students' journal entries. Because our focus is on describing data, we included only excerpts that refer specifically to the students' descriptions of the data or data display. Most students' entries were more complete, including statements in which they compared aspects of the data.

Larry: No teachers have three pets.

Terrance: There's four of them that has three pets.

Janet: The scales for the pets go up by one. The scales for the teachers go up by two. ... No teachers have five pets.

Marcello: The number of pets range from zero to six. The number of teachers range from zero to twelve.

Larry's response characterizes idiosyncratic thinking because he misreads the graph, thinking that the bars represent the number of pets rather than the number of teachers. Terrance seems to be reading the graph correctly, although he does not explicitly state the units of the data values completely. This behavior is typical of a student at the transitional level of thinking. Indicating quantitative thinking, Janet showed an awareness of the display features in her comment about the scales. She also described a selected part of the data. Marcello, from a more analytical perspective, took a broad approach to describing the data. The scale for the number of teachers went from 0 to 16, but Marcello recognized that the range of data values (number of teachers) went to 12.

After students wrote in their journals, the teacher started a whole-class discussion about the graph. This discussion addressed students' misinterpretations of the display. The teacher then launched into the third part of the task by telling the students that the gym teacher was absent when the survey was done. She asked the students to predict, from the data, how many pets he has. As the teacher circulated around the room, the students discussed this question with their partners and wrote their ideas in their journals. The following excerpt is from the ensuing whole-class discussion.

Teacher: How many pets did you predict that the gym teacher would have? Tell us about your prediction, Adam and Eric.

Eric: We got 1.85. First, there were 10 teachers with no pets, then 13 teachers with 1 pet, 12 teachers with 2 pets, 4 teachers with 3 pets, 7 teachers with 4 pets, and 2 teachers with 6 pets. So we added all that up and got 48 for the teachers and 89 pets. We divided 89 by 48 and got our answer.

Janet: [*interrupting*] How can you have one-point-eight-five pets? It doesn't make sense for a prediction.

Eric: It's the average.

Teacher:	Well, Janet, what did you and Derrick predict?
Janet:	Since 13 teachers have 1 pet and 12 teachers have 2 pets, that means most of the teachers have one or two pets. Also, there are teachers that have more than two pets, so we figured he would probably have 2 pets—not 1.85; that doesn't make sense.
Teacher:	Eric and Adam, what do you think about that?
Adam:	Well, 1.85 is close to 2, so if we rounded, we would be at two, also. [*Some discussion follows about computing averages and interpreting the meanings of the numbers in the context of the data.*]
Teacher:	Who had a different way of thinking about it? What were some other predictions?
Larry:	We know our gym teacher also coaches football because my brother is on the team and he's always gone for practice or a game. So, we figured he wouldn't have time for a pet. Our prediction is zero. [*Class discussion ensues about this prediction and how students sometimes use their personal knowledge of a context when analyzing data.*]
Teacher:	I noticed that a couple of you predicted the gym teacher would have 5 pets. How did your analysis of the data lead to that prediction?
Karen:	We thought that at first because the graph showed no teachers had 5 pets, so we thought that he might have 5 and that would fill in all the numbers of pets. But now I think it should be 2, like Janet said.

In this vignette, the students' responses represent all four levels of analyzing data. Characteristic of an idiosyncratic level of thinking, Larry and his partner do not use the given data as a basis for their prediction. Instead, they remain tied to the context of the situation and what they know about the gym teacher. Students' thinking at this level often focuses on irrelevant issues or, as here, focuses solely on aspects of the situation without reference to the actual data. Karen's initial prediction, that the gym teacher would have five pets because the data did not represent that number of pets, is characteristic of students' reasoning at the transitional level. There is some indication that Karen had, in fact, interpreted the data to determine that no teachers in the survey had five pets. However, assuming that another teacher would have five pets simply because that data value had not yet been represented is not reasonable. Her prediction is based more on a visual inspection of the data rather than on an interpretation of the data. One might wonder how she would have responded had there been two gaps in the data or no gaps at all. From a different perspective, Eric and Adam did attend to the data as a whole when they computed the average number of pets. However, they focused on the data as a set of numbers, with no reference to the context. Doing so is characteristic of student reasoning at a quantitative level in that the student uses the data but does not necessarily relate it to the context. A student reasoning at the analytical level incorporates both the data and the context to make reasonable inferences. This approach is evident in Janet's comment that 1.85 pets does not make sense as a prediction for the number of pets the gym teacher would have. Her own prediction was grounded in an interpretation of the data set as a whole. She recognized that "most" of the teachers in the survey have one or two pets, and she took into account the other data values in making a prediction of two pets rather than one.

Connections to practice

These two vignettes illustrate the different levels of statistical thinking that we have seen when working with students in inclusion classrooms at the elementary and middle school grades. What has this to do with instructional practice? Why would a teacher want to identify students' thinking in the context of levels? Our research has shown that when teachers understand the characteristics of student thinking at each level of the statistical thinking framework, they are well equipped to assess students' thinking and, from that assessment, develop learning goals aimed at pushing students' thinking to the next level. For example, in vignette 1, Rosa and Daya constructed a line plot that illustrates a transitional level of thinking for the statistical process of representing data. Their display was not representative of the data; that is, the students did not include on their plot pets that did not fit their categories (garter snakes, iguanas, and chameleons). This reasoning was typical of the representations they had produced over a series of related tasks. Thus, the teacher considered their work to reflect transitional thinking. Having noted the characteristics for representing data at the quantitative level, the teacher decided that later lessons should include tasks to emphasize the importance of representing *all* data when constructing a display. She also considered ways of grouping the students to increase their exposure to higher levels of thinking.

Our work has shown that the levels of statistical thinking do not necessarily depend on age but rather on the experiences students have in data exploration. Thus, engaging all students, including those with special needs, in a variety of activities related to describing, organizing, reducing, representing, and analyzing data is important. For describing data, students should explain what a data display represents and what information they can get from the display. This approach allows the teacher to see what features they notice or overlook. With organizing data, students need opportunities to actually organize and reorganize raw data. Listing information in alphabetical order or from least to greatest can yield some insight to the data, but reorganizing the data into groups or clusters can offer more insights.

Many ways to represent data exist that are not intuitive to students. Box plots and scatterplots such as those in figure 9.7 are useful representations for comparing or analyzing data.

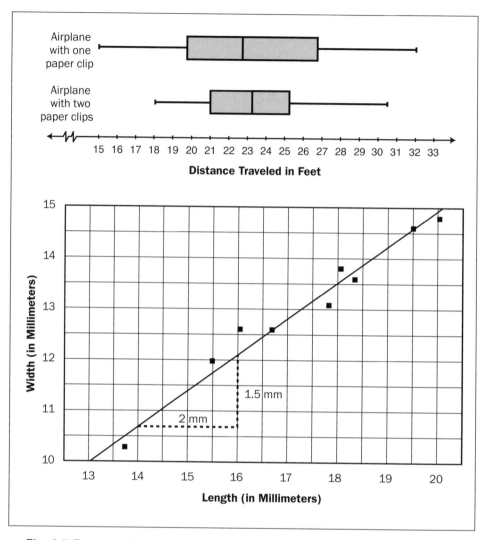

Fig. 9.7. Box plot (upper panel) and scatterplot (lower panel), both from NCTM 2000 (pp. 250 and 252, respectively)

Yet, these are not natural representations that students would consider. Teachers should try to present situations in which representing data in, say, a bar graph is not sufficient for examining the data. For example, representing the categorical data of the number of students who are light, medium, or heavy sleepers in a bar graph would be appropriate. However, a bar graph would not be an appropriate way to display the growth of a plant over time. In analyzing data, students need opportunities to examine data in a wide variety of familiar contexts. Students need opportunities to make predictions with data and to compare multiple data sets. Comparing unequal data sets, such as home runs hit by left-handed batters versus right-handed batters, can help students see the need to use proportional reasoning when analyzing data.

According to the *Principles and Standards* (NCTM 2000), "technology offers teachers options for adapting instruction to special student needs" (p. 25). It can also be useful in developing students' statistical understanding and intuitions. Computers and calculator programs can alleviate some of the toil of generating data displays, thus allowing students to focus on other statistical ideas. However, "technology should not be used as a replacement for basic understandings and intuitions; rather, it . . . should be used to foster those understandings and intuitions" (NCTM 2000, p. 25). Spreadsheet programs can generate a variety of data displays for a set of data. However, the program does not determine the appropriateness of the display or the predetermined scales. The student still needs to understand which displays are appropriate for the data and to understand how to adjust the scales appropriately for the data or for analysis.

Data explorations engage students in meaningful interpretation and analysis of data. The statistical framework in this chapter furnishes a scaffold for examining and developing students' thinking. If students are to develop statistical thinking across the levels, idiosyncratic to analytical, they must be involved in instructional activities that are more open ended and challenging than the typical textbook lesson that instructs them to construct a particular graph or table and then answer factual questions from reading the display.

Connections to families/caregivers

The abundance of data in newspapers and magazines affords many opportunities for parents and caregivers to point out various ways of representing data and to talk about what the data mean. For example, advertisements may contain misleading or product-biased data. Sometimes newspapers design graphs to capture the eye rather than represent data accurately. These contexts can offer opportunities to apply mathematics as well as stimulate conversations about data through questions such as "What can you learn from this display?" "What comparison can you make from the display?" "What predictions can you make from the data?" and "Do you think the data are misleading or biased?" Discussions that result from these questions familiarize students with the variety of ways that entities display data and use data to communicate ideas.

Probability framework

Five key concepts characterize the complexities of probabilistic reasoning: sample space, probability of an event, probability comparisons, conditional probability, and independence. These concepts come from the substantial body of research (e.g., Shaughnessy 1992) on students' probabilistic thinking. This section defines and illustrates each concept.

The concept of *sample space* is fundamental to all probability reasoning. It involves identifying all possible outcomes of an experiment. Initially students explore one-stage experiments, such as tossing a coin. This example has two possible outcomes, heads and tails. Later students investigate two-stage experiments, such as tossing two coins. The sample space for tossing two coins consists of four outcomes: heads–heads, tails–tails, heads–tails, and tails–heads.

The *probability of an event* is based on an analysis of the sample space and uses symmetry, number, or simple geometric measures to determine the likelihood of an event. A student's ability to identify and justify which of two or more events is most likely or least likely to occur demonstrates an understanding of the probability of an event. For example, when playing a game that uses a spinner equally partitioned into five sections, three colored red and two colored blue, a student who understands probability might explain that she has greater likelihood of spinning red because three-fifths of the spinner is red and only two-fifths is blue.

Probability comparisons require students to determine which of two probability situations is more likely to produce a specified (target) event or whether the two situations offer the same chance for the target event. Consider a game that uses the following two spinners:

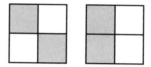

A student understanding probability comparisons can explain that both spinners offer the same chance for spinning red (shown here shaded) because in both cases, half the spinner is colored red.

Conditional probability involves recognizing whether the occurrence of one event has changed the probability of another event. In the following classroom excerpt, Max's explanation exhibits an understanding of conditional probability: Ms. Wilson was selecting three students for classroom jobs by drawing names from a hat. Five students' names were in the hat: Ben, Max, Ann, Patty, and Sara. The teacher selected Ann's name for the first job and removed her name from the hat. Max said, "Now we have a better chance of drawing a boy's name." Ms. Wilson asked him to explain what he meant, and Max replied, "Well at first the boys only had a two-out-of-five chance of being drawn; now it's 50–50: two girls and two boys." Max realized that removing Ann's name from the hat changed the probabilities of subsequent draws.

The concept of *independence* requires students to recognize events for which the occurrence of a particular event does not change the probability of another. The following coin-toss experiment illustrates. *A fair coin has been tossed four times, generating these outcomes: heads, tails, tails, tails. Does heads have a better chance than tails on the next toss?* A student who understands the independence of these events will respond that heads and tails have the same chance of being tossed. The chance of tossing heads is 50 percent regardless of the outcome of the previous toss. A student who does not recognize independence in this situation might respond that heads has a better chance because tails has come up the last three times and so heads being next is more likely.

These five probability concepts address the fundamental ideas in probability that students should develop across the grade levels. The probability thinking framework in figure 9.8 (modified from Jones, Thornton, et al. 1999) characterizes the development of students' thinking about these concepts across the levels, from subjective to numerical. It describes patterns of growth in probabilistic thinking that can help teachers understand the diversity of thinking that students engaged in probability problem situations typically exhibit.

Probability thinking framework

	Level 1, subjective	Level 2, transitional	Level 3, quantitative	Level 4, numerical
Sample space When listing the sample space, the student lists . . .	An incomplete set of outcomes for a one-stage experiment	A complete set of outcomes for a one-stage experiment and sometimes lists a complete set for a two-stage experiment	The outcomes of a two-stage experiment consistently	A complete listing of the outcomes for two- and three-stage experiments
Probability of event When determining the probability of an event, the student predicts most or least likely events . . .	On the basis of subjective judgments and recognizes certain and impossible events	On the basis of quantitative judgments but may revert to subjective judgments	On the basis of quantitative judgments	For one- and simple two-stage events and assigns a numerical probability to an event
Probability comparisons When comparing the probabilities of events, the student . . .	Uses subjective judgments and cannot distinguish between "fair" and "unfair" probability situations	Uses quantitative judgments, not always correctly, and begins to distinguish between "fair" and "unfair" probability situations	Uses valid quantitative reasoning and uses such reasoning to distinguish between "fair" and "unfair" probability situations	Assigns numerical probabilities to events to make a valid comparison
Conditional probability When exploring conditional probability situations, the student . . .	Uses subjective reasoning when interpreting without-replacement situations	Recognizes that the probability of *some* events changes in a without-replacement situation; however, the recognition is incomplete and is usually restricted to events that have previously occurred	Recognizes that the probability of *all* events changes in a without-replacement situation and can quantify changing probabilities in a without-replacement situation	Uses numerical reasoning to compare the probability before and after each trial in without-replacement situations
Independence When considering the independence of events, the student . . .	Has a predisposition to consider that consecutive events are always related and that one can control the outcome of an experiment	Begins to recognize that consecutive events may be related or unrelated and may revert to subjective reasoning	Recognizes when the outcome of the first event does or does not influence the outcome of the second event	Distinguishes between independent and dependent events and can assign numerical probabilities to the events

Fig. 9.8. Probability thinking framework

Examples of students' thinking in probability

The following vignette illustrates the four levels of students' thinking for three of the probability concepts: sample space, probability of an event, and probability comparisons. In this scenario, elementary-level students engaged in an activity called "which spinner?" (See fig. 9.9.) As part of the activity, students worked in pairs to play the "race home" game. The teacher instructed them to record the result of each spin for each game they played. Once students had played the game several times, the teacher started a whole-class discussion. The teacher asked the first two questions as she circulated through the classroom while students were playing the game; she posed the third question during the whole-class discussion. For each question, we present selected students' responses to illustrate the different levels of thinking that the third-grade students exhibited.

Question 1: What colors could come up on spinner 1?

Elizabeth: Red.

Inez: Red and white. But red's my favorite, so it's going to be red.

Kerry: Just red and white.

Corey: Red and white, but they don't have the same chance.

Question 1 addresses the probability process of sample space, specifically for a one-stage experiment. Elizabeth's response reflects a subjective level of thinking: she focused on only one possible outcome, red, and seemed to ignore the possibility of spinning white. Students who reason subjectively about sample space do not list the complete set of outcomes for a one-stage experiment. Inez's response was typical of students exhibiting a transitional level of probabilistic reasoning. She correctly listed the outcomes of the spinner but displayed elements of subjective thinking by predicting a specific outcome, red, because it is her favorite color. In contrast, Kerry and Cory accurately listed the outcomes for spinner 1, a response that would be characteristic of students thinking at the quantitative and numerical levels. However, students' responses to two-stage probability situations distinguish thinking at these levels.

To illustrate the differences between these levels of thinking, figure 9.10 presents a two-stage probability situation from a later lesson. When the teacher asked her to determine the possible sums for spinning two spinners, Kerry eventually listed all the outcomes. Her response reflects thinking at the quantitative level: her strategy for generating outcomes was not necessarily systematic. Corey, however, used a systematic strategy that would be effective for probability situations involving multiple stages. Students like Elizabeth and Inez, whose responses reflected thinking at the idiosyncratic and transitional levels, could not list all the outcomes in this two-stage probability situation.

Question 2: Why does red have the better chance on spinner 1?

Elizabeth: Because everyone likes red.

Which Spinner?

Problem Task: Each player picks one chip, either red or white. Put your chip in the start box of the Race Home Game Board.

Which spinner is better for your color?

> **Race Home Game**
>
> Pick red or white. Put chips on start.
>
> Take turns with Spinner 1.
> If red, move one space.
> If white, move one space.
> Play again. Use Spinner 2.

Spinner 1   Spinner 2

Talk About It: Did the game turn out as you said?

Think and Tell: What spinner is better for the red chip? The white chip? Why?

Fig. 9.9. Which spinner?

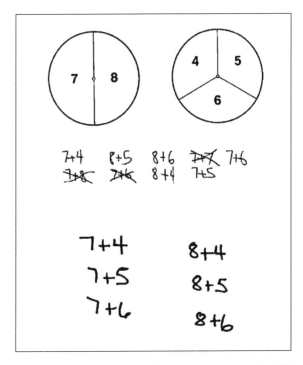

Fig. 9.10. Responses for two-stage probability
situation (top, Kerry; bottom, Corey)

Inez: It's [*the spinner*] mostly red; it's also where I'd start the spinner.

Kerry: There are three red pieces and one white one.

Corey: Red's chance is three out of four, and white's is only one out of four.

This question focuses on determining the probability of an event. Elizabeth's response is typical of subjective thinking: she relied on personal preference (i.e., favorite color) rather than the distribution of the sample space to explain the greater probability for red on spinner 1. In contrast, Inez's response exhibited quantitative reasoning about the situation, although she reverted to subjective thinking for how she would set up the spinner. Her thinking reflects a transitional level of probabilistic thinking. Kerry's use of numbers to quantify the situation is typical of a student using quantitative reasoning. Her response differs from Corey's in that she did not consider the results relative to the whole spinner. Using proportional reasoning in determining the probability of an event indicates thinking at the analytical level.

Question 3: Which spinner is better if you are playing the red chip?

Elizabeth: It doesn't matter; they both have red. Red always wins.

Inez: The first spinner has more red.

Kerry: Spinner 1 because there are three red parts [*on the first spinner*] instead of one [*red part on the other spinner*].

Corey: On spinner 1, it's three-fourths red. On spinner 2, it's one-fourth red. So you have a better chance of spinning red with spinner 1.

Determining which spinner is better for red addresses the process of probability comparisons. Elizabeth's response reflects subjective thinking: she shows no regard for the fairness of the two spinners. In her mind, both spinners have red and red always wins, so either spinner is OK. Elizabeth has not yet developed an understanding of the differences in using spinner 1 or spinner 2. This finding is consistent with the level of thinking she exhibited on questions 1 and 2. Inez's response indicates that she has begun to distinguish the fairness of the two spinners, although she does not specifically compare the probability of spinning red for each spinner. This approach represents a transitional level of probabilistic thinking. Kerry's response reflects quantitative thinking because she compares the number of red parts on each spinner to validate her reasoning. At a numerical level of thinking, students typically assign a numerical probability to each event (i.e., spinning red). Corey's response exemplifies this level of probabilistic thinking by referring to the fraction of red on each spinner.

Connections to practice

The preceding vignette comes from an instructional program in probability that we conducted in an elementary-level inclusion classroom. Our work with these students, and others in the upper elementary and middle school grades (Jones, Langrall, et al. 1999; Jones, Langrall, et al. 1997; Jones, Thornton, et al. 1999), has shown that many students are limited in their intuitive, informal understandings of probability concepts and some hold deeply seated misconceptions about probability. However, almost all students show growth in probabilistic thinking after instruction. The probability thinking framework guided our instructional programs in that we assessed students' thinking and designed instructional activities on the basis of our knowledge of the patterns of growth that the framework described. We selected probability tasks that addressed one or more of the four probability concepts and that were accessible to students functioning at different levels of thinking.

The "which spinner?" activity in the vignette is an example of such a task. Regardless of their levels of thinking, all students in the class could play the race-home game and participate in the whole-class discussion. As the teacher circulated through the classroom while the students played, she posed different questions to students according to her assessment of their understanding of the probability concepts involved in this task (i.e., sample space, probability of an event, and probability comparisons). Her questions were specifically designed to challenge students' thinking. For example, knowing that Elizabeth tended to make subjective judgments about probability situations, the teacher asked her questions about the certainty and impossibility of spinning red or white. She also asked Elizabeth to consider what had happened with the spinner when she and her partner were playing the game. In contrast, the teacher asked whether Kerry (whose responses typically reflected quantitative thinking) could make a fair spinner for the game by using three colors. The teacher hoped that this task would encourage Kerry to use numerical probabilities to justify that one could construct a three-color spinner in a way that was fair. The teacher's knowledge of the probability framework and the characteristics of students' thinking at different levels guided her in posing questions that promoted thinking at higher levels. She was aware, however, that students would need many experiences with these concepts to fully develop their understanding of probability.

We have found several instructional strategies helpful in promoting development of students' probabilistic thinking. Most important, students must be engaged in problem situations that challenge their current levels of thinking. They need to be confronted with problems that involve a level of uncertainty or element of chance and whose solutions call for collecting and interpreting data. These problems need to be complex enough to create a need for modeling the situation, yet remain accessible to students at different levels of thinking. Students need to be familiar with and use a variety of probability models (e.g., dice, spinners, computer simulations) when conducting experiments and determining *what is fair*. They need experiences analyzing situations that use both small and large samples of data. In summary, students need opportunities to predict probabilities, conduct experiments, collect data, and analyze the findings to solve probability problems. This type of problem solving, based on prediction, experimentation, data collection, and analysis, has enriched students' thinking and challenged some strongly held misconceptions. Ongoing experiences such as these develop students' probabilistic thinking across the levels, subjective to numerical.

Connections to families and caregivers

Families and caregivers have many opportunities to nurture their children's understanding of probability. The most common is through games that involve elements of chance. For example, board games that use spinners or dice naturally evoke discussions about the likelihood of certain outcomes and what is fair. Similarly, card games introduce students to notions of randomness and predictability. Teachers should encourage families and caregivers to talk with their children about situations in which outcomes or events are certain, likely, unlikely, or impossible. Through such experiences, students begin to develop intuitions about probability and to construct informal understandings of probability concepts such as sample space, probability of an event, probability comparisons, conditional probability, and independence.

Reflections: Using Frameworks to Foster the Growth and Development of Students' Thinking

Cognitive frameworks such as those we described here offer insights into students' thinking that can assist teachers in fostering the growth and development of students' reasoning. Our research has shown that such background knowledge is helpful in planning, implementing, and evaluating instruction. Teachers who understand the general patterns of growth of students' thinking are better equipped to plan lessons that include tasks and questions aimed at different levels of thinking. Thus, they are more likely to meet the diverse needs of students in their classrooms.

During instruction, teachers can refer to thinking-level descriptors as they assess or interpret students' responses to problem tasks. Again, this approach enables teachers to better accommodate diversity in students' reasoning and guides them in posing questions that move students toward more mature levels of thinking. Finally, cognitive frameworks afford helpful benchmarks for assessing growth in students' thinking and evaluating the effectiveness of instruction. Teachers can use framework descriptors to profile the development of students' thinking to help them continue the cycle of planning, implementing, and evaluating instruction.

When teachers use cognitive frameworks as the basis for instructional decisions, they are more likely to maintain high standards and engage all students in challenging mathematics. Rather than "watering down" the curriculum to accommodate the diversity among students' levels of thinking, teachers are focused on promoting more advanced levels of thinking. Moreover, cognitive frameworks acquaint teachers with the big ideas of a mathematical domain, especially ones such as data analysis and probability, that they may not be familiar with.

Reflections on Teaching Data Analysis and Probability: Questions for Discussion

1. Students should have multiple learning experiences with a variety of data representations and probability models. What accommodations or special supports might you need to offer for students with different needs?

2. Critique a probability task you have used with students. How is it accessible to students who are reasoning at different cognitive levels? How could you modify the task to make it more accessible?

3. Consider the diversity of thinking that the students in vignette 1, in the data analysis section of the chapter, reflected. What would be an appropriate *next lesson* for this class? Why?

4. We urge teachers to avoid using the thinking levels from our frameworks as labels for students. Similarly, the thinking-level descriptors are not objectives to be mastered to move to higher levels of thinking. How then can you use these cognitive frameworks to assist you in planning instruction and assessing student thinking?

References

Biggs, John B., and Kevin F. Collis. "Multimodal Learning and Quality of Intelligent Behavior." In *Intelligence: Reconceptualization and Measurement*, edited by Helga A. H. Rowe, pp. 57–66. London: Psychology Press, 1991.

Council of Chief State School Officers (CCSSO). *Common Core State Standards*. Washington, D.C.: CCSSO, 2010. www.corestandards.org.

Curcio, Frances R. "Comprehension of Mathematical Relationships Expressed in Graphs." *Journal for Research in Mathematics Education* 18 (November 1987): 382–93.

Franklin, Christine, Gary Kader, Denise Mewborn, Jerry Moreno, Roxy Peck, Mike Perry, and Richard Scheaffer. Guidelines for Assessment and Instruction in Statistics Education (GAISE) Report: A Pre-K–12 Curriculum Framework. Alexandria, Va.: American Statistical Association, 2007. www.amstat.org/education/gaise.

Friel, Susan N., Frances R. Curcio, and George W. Bright. "Making Sense of Graphs: Critical Factors Influencing Comprehension and Instructional Implications." *Journal for Research in Mathematics Education* 32 (March 2001): 124–58.

Jones, Graham A., Cynthia W. Langrall, Carol A. Thornton, and Timothy A. Mogill. "A Framework for Assessing and Nurturing Young Children's Thinking In Probability." *Educational Studies in Mathematics* 32 (February 1997): 101–25.

———. "Students' Probabilistic Thinking in Instruction." *Journal for Research in Mathematics Education* 30 (November 1999): 487–519.

Jones, Graham A., Carol A. Thornton, Cynthia W. Langrall, Edward S. Mooney, Bob Perry, and Ian J. Putt. "A Framework for Characterizing Children's Statistical Thinking." *Mathematical Thinking and Learning* 2 (October 2000): 269–307.

Jones, Graham A., Cynthia W. Langrall, Carol A. Thornton, Edward S. Mooney, Arsalan Wares, Marion R. Jones, Bob Perry, Ian J. Putt, and Steven Nisbet. "Using Students' Statistical Thinking to Inform Instruction." *Journal of Mathematical Behavior* 20, no. 1 (First Quarter 2001): 109–44.

Jones, Graham A., Carol A. Thornton, Cynthia W. Langrall, and James E. Tarr. "Understanding Students' Probability Reasoning." In *Developing Mathematical Reasoning in Grades K–12*, 1999 Yearbook of the National Council of Teachers of Mathematics (NTCM), edited by Lee V. Stiff and Frances R. Curcio, pp. 146–55. Reston, Va.: NCTM, 1999.

Kilpatrick, Jeremy, Jane Swafford, and Bradford Findell, eds. *Adding It Up: Helping Children Learn Mathematics*. Washington, D.C.: National Academies Press, 2001.

Mooney, Edward S. "A Framework for Characterizing Middle School Students' Statistical Thinking." *Mathematical Thinking and Learning* 4 (January 2002): 23–63.

National Council of Teachers of Mathematics (NCTM). *Principles and Standards for School Mathematics*. Reston, Va.: NCTM, 2000.

———. *Curriculum Focal Points for Prekindergarten through Grade 8 Mathematics: A Quest for Coherence*. Reston, Va.: NCTM, 2006.

Shaughnessy, J. Michael. "Research in Probability and Statistics: Reflections and Directions." In *Handbook of Research on Mathematics Teaching and Learning*, edited by Douglas A. Grouws, pp. 465–94. New York: Macmillan, 1992.

Shaughnessy, J. Michael, Joan Garfield, and Brian Greer. "Data Handling." In *International Handbook of Mathematics Education*, edited by Alan J. Bishop, M. A. (Ken) Clements, Christine Keitel, Jeremy Kilpatrick, and Colette Laborde, part 1, pp. 205–37. Dordrecht, Netherlands: Kluwer Academic Publishers, 1996.

Index